SpringerBriefs in Electrical and Computer Engineering

W0090676

For further volumes:
http://www.springer.com/series/10059

Richard Doornbos · Sjir van Loo
Editors

From Scientific Instrument to Industrial Machine

Coping with Architectural Stress in Embedded Systems

 Springer

Editors
Richard Doornbos
Research Fellow
Embedded Systems Institute
Den Dolech 2
5600 MB
Eindhoven
The Netherlands

Sjir van Loo
Senior Research Fellow
Embedded Systems Institute
Den Dolech 2
5600 MB
Eindhoven
The Netherlands

ISSN 2191-8112 ISSN 2191-8120 (electronic)
ISBN 978-94-007-4146-1 ISBN 978-94-007-4147-8 (eBook)
DOI 10.1007/978-94-007-4147-8
Springer Dordrecht Heidelberg New York London

Library of Congress Control Number: 2012935675

Printed on acid-free paper

Springer is part of Springer Science+Business Media (www.springer.com)

Foreword

"There is plenty of room at the bottom" is the title of a famous lecture in which Richard Feynman, Professor of Physics at several US universities, challenged electron microscopy to improve its resolving power. "It would be really easy to understand properties of materials, we only have to look at it and see where the atoms are. Is there no way to make electron microscopes powerful enough to do this?" This is exactly what FEI has been doing over the last decades: improving the resolving power from about 2 nm in 1960 to the current 0.05 nm; and improving detection capabilities to the point where we can determine the position of individual atoms in a thin sliver of material.

About 5 years ago, however, we realised there was another challenge to face. Our fantastic instruments were highly inflexible and really only suited to work in a research laboratory. How could we make them so they would also be suitable for more demanding environments in several industries. And capable of being operated by persons whose primary interest is not so much in microscopy per se, but in the results they could get. A typical market where this was already happening was the electronics industry. To develop semiconductor processes and designs, and to ramp-up novel processes in the factory and analyse failures, they needed our very best high-end transmission electron microscopes (TEMs). They were interested in automatic applications, in reliable performance and reproducible results. And this trend did not just have an industrial driver. Our life-science customers, even in universities, started making similar demands for their microscopes: higher throughput, quality and reliability of data, etc. At the same time, we realised this challenge could not be solved by building dedicated tools which would only fit one particular application: cost and development efforts would be prohibitive.

In our first contacts with the Embedded Systems Institute, we put the challenge to be investigated in what became the Condor project as follows: "How can we transform our electron microscopes from a research instrument, used by specialists, to an industrial-grade, person-independent measurement tool, without losing flexibility?" Together with ESI we defined a series of subprojects to investigate this challenge from different angles, all on the general foundation of model-based design. Subjects ranged from automatic alignments (image-based), via motion-control, scanning strategies and magnetic field control, towards novel system architecture concepts and implementations. We found partners in the universities of Antwerp, Leuven, Delft and Eindhoven, and in the company Technolution. Some groups were already familiar with electron microscopy; others needed to invest in understanding this field. Finding the right students was no easy task either, but about one year after the project was approved, the new PhDs and postdocs, together with a team from FEI and ESI, started to work on their subjects.

Looking back on the project, we see several types of results from which the various partners profit in different ways. There has been an important scientific output in terms of peer-reviewed papers, conference contributions and PhD theses. At the same time, it has been an interesting experience for some of the groups and individuals to perform scientific investigations in a more industrial environment ("Industry as a lab"), and sometimes really at (or just beyond) the boundary of their original competences. For ESI, there has been a major knowledge build-up in the field of architecture for electron microscopy, or scientific instruments in general. For FEI, the first result has been to get an outside view from the ESI and university partners into the technology and architectures we develop in-house. Some initial investigations from partners busted a few myths which originated from too narrow a view, and not enough time spent on properly investigating some aspects "which always worked this way". In other cases potential solutions were investigated that FEI would originally have regarded as far-fetched, like using field-sensing probes for fast control of lens fields. Some investigations yielded results that could easily be implemented, like the application of a Nelder-Mead optimisation algorithm for beam control and alignment. And of course there are the results yet to be implemented in new product developments. This book gives an excellent overview of the more tangible results which this project achieved.

Did FEI reach its objective: did we transform our instruments into flexible but more industrial-grade tools? It has not always been easy to see this through the very dynamic business environment at FEI, but the long-term strategy in this direction is even stronger than when we started. To really transform our architectures and systems is a long and difficult process. However, it is really happening, and the Condor project helped to overcome some of the first hurdles!

Eindhoven   Dr. Frank de Jong
15 December 2011   Director Research and Technology FEI

Preface

In 2007, the Condor project was set up as a first step to answering the question "how can we adapt the current system architecture of an electron microscope to obtain a predictable and automated system that can be used in industry?" While the question was posed with a specific type of system in mind, namely one of FEI Company's electron microscopes, both the question and the project's results are equally valid for other systems.

This book focuses on the main challenges of the Condor project, outlining these and summarising the research carried out across the various routes taken towards solutions. The project showed that the issue of "architectural stress" is of great importance to many industrial businesses. Turning a purpose-built precision-critical system, such as the classical FEI electron microscopes, into a system that is more flexible, predictable and more easily adapted to changing circumstances is rapidly becoming a desirable pathway to evolving new systems.

This book is the seventh in our industry-as-laboratory projects series.[1] These large 5-year projects are put together and led by the Embedded Systems Institute (ESI), in close collaboration with its industrial and academic partners.

In addition to ESI, Condor involved a consortium of industrial and academic partners. The industrial partners were FEI Company, the carrying industrial partner, and Technolution. The academic partners were the Eindhoven University of Technology, Delft University of Technology, Katholieke Universiteit Leuven and University of Antwerp.

---

[1] Books about earlier industry-as-laboratory projects—Falcon, Boderc, Tangram, Ideals, Trader and Darwin—can be found on ESI's website: http://www.esi.nl/knowledge-transfer/publications/books.

I would like to thank all partners and individuals, researchers and managers, companies and academia alike, involved in the Condor project. Through individual efforts and splendid teamwork, they contributed to the project's success. It is now time to pass on the experience gained and results achieved to a wider industrial and academic audience.

Eindhoven   Prof. dr. ir. Boudewijn Haverkort
February 2012   Scientific Director and Chair
                                                             Embedded Systems Institute

Acknowledgments

From scientific instrument to industrial machine is a result of the Condor project conducted under the responsibility of the Embedded Systems Institute with FEI Company as the carrying industrial partner. This project is partially supported by the Netherlands Ministry of Economic Affairs under the Embedded Systems Institute (BSIK03021) program.

We gratefully acknowledge the cooperation of the FEI employees throughout the 5-year project. Electron microscopy is an intriguing field. It combines cutting-edge technology with a high system complexity and has led to many Nobel prize-level results. Together with our academic partner's research and the vast experience and knowledge in electron microscopy of FEI's employees, especially Seyno Sluyterman, Mart Bierhoff and Auke van Balen, we performed our research on building complex systems. This formed the basis for *From scientific instrument to industrial machine*, a book aimed at system architects, engineers and other practitioners working on the challenges of creating complex high-tech systems.

We would like to thank the employees of our partners (Embedded Systems Institute, FEI Company, Technolution, University of Antwerp, University of Leuven, Delft University of Technology, Eindhoven University of Technology) for contributing to this book.

We would also like to thank the reviewers from FEI Company, Daimler, Delphino Consultancy and the Embedded Systems Institute.

Finally we express our special thanks to the Condor project leader Jan Schuddemat of the Embedded Systems Institute.

Contents

Part IV Conclusion

Part I
Introduction

Chapter 1
The Endeavour

Over the last 5 years, the editors of this book together with a project leader, a large multi-disciplinary team of researchers and industrial engineers, and support staff were involved in an applied research project called Condor. The Condor project investigated how to adapt a highly specialised, complex system to meet new market requirements. Taking a transmission electron microscope (TEM) as a prime example of a complex system, the project focused on:

- Investigating methods to automatically find the right settings for the electron microscope and relieve the operator from the burden of correcting for drift
- Finding new architectural concepts and constructs to support the change, and
- Combining these two into a working proof of concept, as a convincing demonstrator for our research.

In addressing these problems, we discovered that such systems suffer from *'architectural stress'*—the inability of a system to be easily adapted to new applications. This book is the result of our research and describes our investigation of 'architectural stress' and the challenges we overcame to make the most of the engineering and scientific qualities, and ingenuity of all team members involved.

This project was a close cooperation between industry and research. Industry provided the problem within a business context and research took up the specific challenges and provided concept solutions. As an intermediate between these parties, we fulfilled the role of obtaining a full understanding of the problem within the business context. We then sub-divided it into partial problem statements, to be worked on by the various research teams. Also, together with the researchers and industrial engineers, we integrated the partial solutions into an operational system as a proof of concept to obtain a first step towards a next generation of products.

This method of working, referred to as industrial or applied research, does not measure its success by the number of scientific papers. Instead it takes the industrial uptake of results and ideas as the main measure of success for the industrial problem owner.

R. Doornbos and S. van Loo (eds.), *From scientific instrument to industrial machine*, SpringerBriefs in Electrical and Computer Engineering, DOI: 10.1007/978-94-007-4147-8_1, © The Author(s) 2012

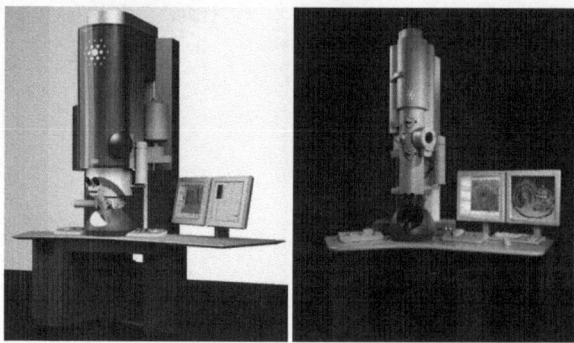

Fig. 1.1 The Titan 80/300 and Tecnai TF20

The Challenge

The industrial partner in the Condor project was FEI Company. FEI manufactures electron microscopes and is the world leader in high-end TEMs with their Titan and Tecnai product families (Fig. 1.1). For many years their key design driver has been to improve the machine's resolution, i.e. to make increasingly smaller structures visible to the eye. This has led to an impressive instrument capable of visualising sub-Ångström details in e.g. atom grids.

The electron microscope is a complex instrument with an electron gun, electromagnetic lenses, a sample holder on a stage, detectors for electrons and X-rays (see Fig. 1.2), and supporting equipment such as a high-voltage generator and a vacuum system.

Electron microscopes are being increasingly used in industrial workflow and lab procedures, where the resolution available today is good enough. These new industrial applications required the design of a new architecture for the TEM system. For instance, there is greater need for flexibility and adaptability to the customer's working practices.

Key drivers for this new generation of electron microscopes are: cost-effective analysis, accurate measurement, availability and ease-of-use. These are significantly different from the traditional drivers of high resolution and image quality. To address these new markets and applications, the inherent limitations of the current systems architecture can hamper, or even block, the introduction of new solutions. This problem is called '*architectural stress*'.

Many industries are going through a comparable phase, e.g. professional printers and medical imaging systems. Being complex machines, like the electron microscope, it is not feasible to redesign them from scratch. Therefore, the only way forward is to build upon the available architecture, changing it and improving parts gradually.

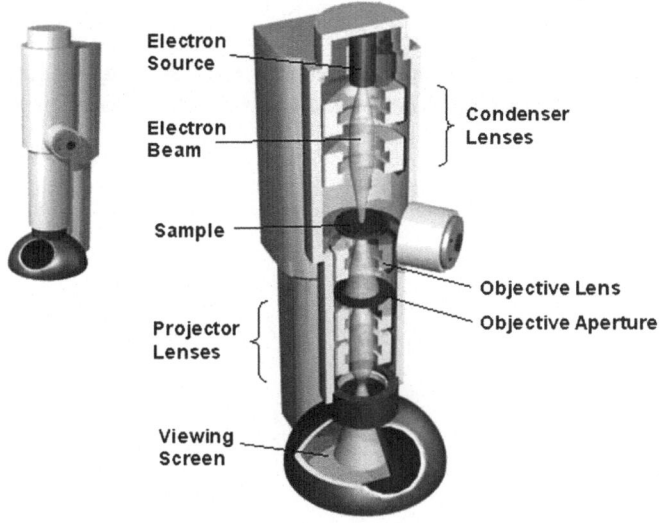

Fig. 1.2 Internals of a transmission electron microscope (*illustration by Kevin Yager*)

The Book

This book is aimed at all systems architects, designers and engineers of complex systems. The goal is to provide an example of how to approach a problem like this, where newly required key qualities conflict with the dominant system qualities from the past.

Consequently, our contribution is to the field of systems engineering. The book does not have a deep scientific nature, and mathematics is avoided as much as possible. The *focus* of the book is a reflection on our endeavour, its results and the problems we encountered. We aim to abstract from the electron microscopy specifics, making it possible for system architects and engineers to mimic our approach when dealing with similar design challenges. So, it does not elaborate on all the ins and outs of architectural stress, but merely illustrates some of the aspects we encountered in the five years of our project.

The book is not just a collection of individual contributions. The editors provided the bones, the structure of the main problem domains and an introduction to these problems. The individual contributions by the project participants form the meat attached to the bones. All the individual contributions used a common approach, describing the high-level problem in the context of the electron microscope; the problem and the solution direction in generic terms; the solution as applied in the context of the electron microscope; and the lessons learned when applying the results in a different context.

We organised the subjects into two main categories: architectural topics, and automation and control topics. The book ends with a discussion on the observations and lessons learned during the Condor project.

Part II
Architecture

Chapter 2
Systems Architecture

Abstract This introductory chapter discusses the important system architectural aspects of a transmission electron microscope and how these are influenced by new market demands. It starts by discussing system-wide architectural considerations and the challenges imposed by current and future industrial markets. Then it looks at how new market demands (e.g. automation, repeatability and ease-of-use) change the architectural key drivers significantly, and therefore lead to *architectural stress* in the current system. Finally, architectural patterns and new design concepts for a next system architecture for the transmission electron microscope are discussed.

Keywords Systems architecture · Electron microscope · Key drivers · System qualities

2.1 Architectural Considerations

Richard Doornbos

Embedded Systems Institute, The Netherlands

Firstly, it is important to understand how the current state-of-the-art electron microscopes can be characterised as a system. In general, one can say that they are precision-critical: overall system performance–in particular resolution–is highly determined by the accuracy and precision of its components. As current TEMs are capable of extreme magnification, e.g. for visualising single atoms, it should be clear that the quality requirements of all critical components are at the cutting edge of what is technically possible.

R. Doornbos and S. van Loo (eds.), *From scientific instrument to industrial machine*, 9
SpringerBriefs in Electrical and Computer Engineering,
DOI: 10.1007/978-94-007-4147-8_2, © The Author(s) 2012

Electron microscopes are long-lived systems, as main components like column, mechanics and electronics remain operational for at least 15 years. Only peripheral components, such as computer hardware and system software, are typically updated on a shorter time scale. A main characteristic of these long-lived systems is that they generally suffer from obsolescence: that components needing to be replaced are already out of production.

Traditionally, electron microscopes are intended for basic and exploratory research in the fields of material sciences, performed in an academic context. To successfully operate these systems takes an expert in physics, usually dedicated for years to a particular system. They will be both knowledgeable about the internals of the microscope, to optimally control its operation, and understand the physics of the electron-specimen interaction and the microscopic application to be investigated.

It is clear from the above that current electron microscope system designs focus on key qualities like resolution and image quality.

Trends

As the resolution of the best modern TEMs (e.g. Titan from FEI Company, introduced in 2005) is in the order of 50–80 pm (1 pm $= 10^{-12}$ m), it is often said that the resolution race is over. There is no strong need to significantly increase the resolution further as the size of the smallest atom lies in the order of 120 pm. Therefore this probably marks the end of current challenges.

Looking at current trends in microscopy applications, we generally see a significant move from human observation towards measurement and quantification. There is a growing need for statistical evidence, requiring repetitive accurate measurements and routine applications.

It is hard to explore these trends in detail, as a very large variety of applications exist or are being developed. The market is also rather fragmented. Typical research applications aim for ultimate and unique results that will lead to scientific recognition. However, industrial applications aim for entirely reliable results in the area of (troubleshooting in) production process development and process quality control.

New *types* of microscopy applications are being developed, enabling the measurement of the smallest signals as computational support increases. The power of modelling and reconstruction techniques in this area is impressive.

In electron microscopes, the pace of development has traditionally been governed by the speed of technological and engineering innovation regarding higher resolution. With the resolution race coming to a halt, things may change towards a more customer driven innovation model.

A New Direction

To understand what the new markets are and what these different customers want from the electron microscope, it is very important to thoroughly analyse the new

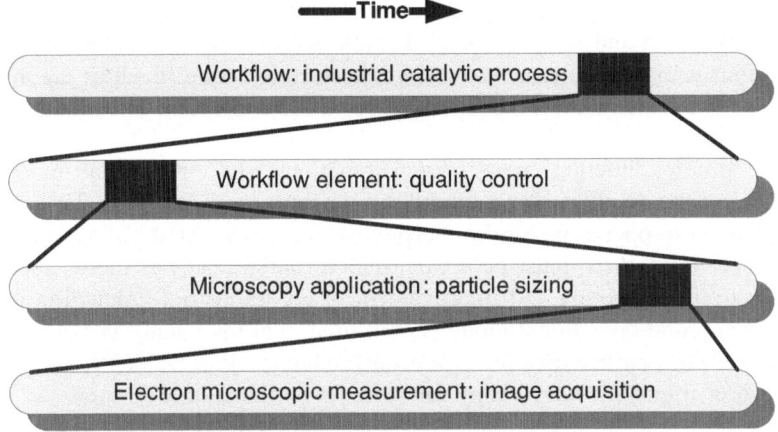

Fig. 2.1 Hierarchy of an (example) industrial usage of electron microscopy

needs and requirements. What is exactly meant by 'routine measurements' and 'automated nanometrology'? The answer will probably be different for each customer, but a common denominator should result.

For a company producing systems in this field it is also important to search for new 'key differentiators' and 'distinctive features', e.g. strong automation, high stability and system predictability, excellent ease-of-use, etc.

Of course, a change of direction like this is likely to have more profound business consequences. Not only may markets change, but different business models may become more profitable. In particular, the need for more automated measurements suggests the role of a *solution provider*. This entails involvement in the measurement problem, provisioning of measurement solutions and reporting support. Activities such as keeping the system operational will often also be part of the solution provider's responsibilities.

An important technical consequence is the embedding of the system into the customer's workflow. This involves the specimen flow, such as sampling, preparation, loading and unloading, disposal, together with the integration of the measurement system into the customer's computational and IT infrastructure.

Investigation of the customer's workflow provides new perspectives on the role and importance of the electron microscope in the entire process. There are new insights in process optimisation with respect to throughput. They show that speed increase in the electron microscopic measurements sometimes has very little effect on overall throughput (see the example in Fig. 2.1).

The new types of applications and their industrial context strongly influence the design of a new electron microscope systems architecture. The architecture should provide more flexibility and adaptability with respect to the customer's way of working. Generalizing over the entire field introduced so far, the new *key drivers* [1] are: cost effective analysis, accurate and precise measurements, and ease-of-use.

Note that these drivers are significantly different from the traditional key drivers of high resolution and image quality. Therefore, addressing the new market and applications using the current systems architecture is very difficult as the inherent limitations hamper, or even block, this process. This problem is called *'stress in the architecture'*.

With a new direction, new system functions and operational modes are required. At the workflow and workflow element level (see Fig. 2.1), we can distinguish two operation modes needed by an anticipated future automated microscope. First, there must be a batch-mode routine analysis mode, especially suited to trouble shooting and improvement of e.g. industrial production process steps. The second operational mode is an in-line, real-time analysis of production process samples, particularly for industrial production process control.

The most important microscopy functions at application-level are:

- Determination of nano-particles' sizes
- Determination of the three dimensional shape of nano-particles or aggregates, and
- Determination of the chemical composition of nano-particles.

To be able to perform these measurements cost-effectively, while eliminating repetitive labour (e.g. manually analysing thousands of particles), automation is essential. We can distinguish several major elements in the measurement process that are eligible for automation: specimen preparation, specimen loading and unloading, imaging by the electron microscope, and image measurement and data interpretation (this may include off-line image processing).

At the electron microscopic measurement level we need new basic system functions. These include automated acquisition of an optimum image (e.g. 'sharp', minimum distortions, sufficient signal-to-noise ratio), automated acquisition of a specified area of the specimen, etc.

The *key system qualities* [1] which should follow from the key drivers can be related to the levels indicated in Fig. 2.1. At the highest level, availability and reliability (note that the electron microscope system is part of the workflow!), and adaptability and flexibility with respect to customer needs are the most important system qualities, as they will lead to cost-effective analyses.

At the microscopy application level, throughput or short time-to-result, and understandability are the most important system qualities, as they lead to cost-effective analyses and address ease-of-use.

At the electron microscopic measurement level, system stability, predictability, precision, reproducibility, and robustness are the most important system qualities, as they lead to result correctness and address ease-of-use.

The consequences for the operator will be significant, as the built-in automation eliminates the need to deeply understand the system to a large extent. Therefore less dedicated education is needed and 'non-PhD operators' can use the system. However, specialised knowledge is still needed to fully exploit the new possibilities, especially on automation scripts, automated application, and experiment design. This signifies the split between a microscope operator and an experiment designer/application specialist. The user-system interaction will also be significantly different: there is less user control of the system's operation, particularly during automated measurement runs. We suggest adding a clear user notification mechanism about the so-called *under the hood* operation (to know what is going on), and about the effects of automation. Effects can include a certain amount of extra sample damage, the extra time needed for system control, and the possibility of irrecoverable errors.

Fortunately, less time will be spent by operators on tuning the system, making more time available for the actual measurement task. Associated with this change we expect that more off-line and remote analysis work will become the norm. The user's trust in the generated data should be maintained. We suggest introducing sufficient feedback and checkpoints in e.g. automated image interpretation, by using partial/user-assisted instead of fully automated image interpretation.

For a system architect responsible for the new family of electron microscopes, this change leads to many complex questions and discussions, and opens up various investigations. Besides participating in finding out what functionality is required, the system architect's role is to provide answers on how to realise these new systems given the design of the existing systems, the current development organisation, and time and cost constraints.

The main technical questions are about which new techniques, methods, or even system architectures are suitable. The system architect is asked to give a concrete answer to the question: how can we adapt (add-on, redesign) current (precision-critical and obsolescent) systems to achieve the new requirements (typically still in vague terms, or in terms of physical components) within given constraints (cost, time, and organisation)?

Concepts and Possible Solutions

In finding answers to the questions above it is important to analyse the current system for bottlenecks and system tensions that arise due to the new requirements. A simple back-of-the-envelope throughput calculation using the current TEM shows that only low numbers of samples per hour can be analysed. This is, for example, mainly due to slow loading and unloading of specimens. Another analysis reveals that automated imaging of a large area (needing multiple images) requires improved stability and the correction of side-effects. These issues are addressed in detail in Part III of this book (Chaps. 5–7).

Finding conceptual solutions is a core task of the system architect. There are few formal approaches or tools to support this. However, by examining the architectural patterns, creativity can be stimulated and steered in the right direction. In the next section, the architectural patterns discovered in current TEMs are discussed and new system concepts for future electron microscopes are suggested.

2.2 Design Concepts for Global Electron Microscope Control

Sjir van Loo

Embedded Systems Institute, The Netherlands

Traditionally, an electron microscope operator often just wants to take a picture in order to visualise certain aspects of the sample at hand. And, as with a photo camera, the operator first has to set the microscope into an operational mode to ensure images fulfil certain quality parameters. The operator may control a relatively large number of settings, many of them while observing the generated image. The visual feedback loop is used to optimise the image until the operator decides that the image is 'good enough'. Tasks the operator performs include inserting the sample into the microscope, setting the magnification, bringing the sample into focus, removing astigmatism from the image, finding a feature of interest on the sample, and so on.[1]

After the operator has set the microscope into a desired setting, without touching any of the controls image quality will slowly degenerate over time. There are 'aging' effects of the sample, caused by bombarding the sample continuously with high energy electrons, and contamination effects. Besides these, most of the effects observed are caused by internal and external physical phenomena. They include relaxation effects of the electrical and mechanical parts, temperature changes in these parts, external pressure changes, and other phenomena that influence the microscope's settings. So the operator has to change the settings continuously to maintain the desired image quality. Essential for this discussion is that on one hand, an operator may actively change the image by adjusting settings like magnification, sample position or electron beam related parameters like energy and spot size, while on the other hand the machine will slowly drift away from its set point by internal and external influences.

Regarding automating microscope usage, the goal is to change the architecture so that the system shows autonomous behaviour. This involves replacing many of the actions normally performed by the operator. At first this implies that a measurement is automated and consequently the user has to be able to write an application program, performing the desired measurements. As an example see the application described in Sect. 5.2. The application programmer's most important task is to program the analysis of the images acquired. As such it is very helpful if the images acquired always have the quality desired. That is, they are 'sufficiently' in focus and astigmatism free, and not, or barely, drifting away. For simplicity, we will assume that 'sufficiently' can be defined and is fixed for all applications. See

---

[1] In doing so the user builds up a mental model of the sample, how it is structured, which parts are elevated, where the interesting parts are, etc. The actual image is often just a reminder to trigger the mental model.

also Sect. 6.3. The system has to perform a number of actions *under the hood*, i.e. invisible to the application programmer.

For the sake of this discussion we will divide the control actions the system can perform into three different groupings:

1. Electron microscope settings
   These are the global application choices for a number of controls that define the way the microscope is used. They include, among others, the mode (e.g. camera or scanning acquisition mode), the magnification, the electron beam parameters like spot size, and many others. Typically these do not change during image acquisition. Settings are assumed to be constant.
2. Position controls
   Several means exist to position the sample in 3D space with respect to the position of the beam. See Chap. 7.
3. Image controls
   These are used to control important image quality parameters such as focus and absence of astigmatism. They cannot be 'set', but have to be optimised based on observation of the images generated. We will limit ourselves to focus and astigmatism.

If changes are made to the electron microscope. settings or the position controls, the image controls may have to be adjusted as the image quality might degrade, e.g. get out of focus. The impact of some changes will be very limited, while for other changes the precise effect is unknown. This suggests a model in which changes in settings and positions can affect the image controls according to a set of simple rules. These could be "after changing the spot size we have no knowledge about the state of focus and astigmatism", "one step change in magnification has a limited effect on focus and no effect on astigmatism", and "a stage move where (move distance × magnification < value) has a limited effect on focus and no effect on astigmatism".

For our architecture these rules are important. Given that we want to provide the user with images of sufficient quality, we need a way of determining if the system needs to perform an action, like focusing the image. So, if an application changes the microscope settings or repositions the sample, the system has to intercept and invoke an appropriate action as defined by the given rules. The system must maintain an invariant, informally defined as providing the application with images of sufficient quality only.

Invariant

For the following it is useful to introduce a set of discrete states for the key characteristics. For focus we come to three states: OK|CLOSE|UNKNOWN. For absence of astigmatism, we have two states: OK|UNKNOWN. It turns out that when combining these only 3 states can result from the above rules: (OK, OK), (CLOSE, OK), (UNKNOWN, UNKNOWN).

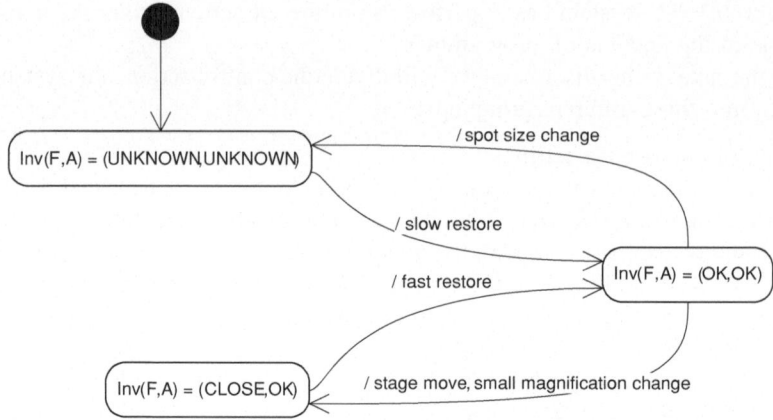

Fig. 2.2 State diagram for the invariant

To be able to acquire a useful image we define that an invariant expression on image controls should hold. The invariant expression is straightforward: (Focus == OK) AND (Absence-of-astigmatism == OK), denoted as Inv(F, A) == (OK, OK). As initially the status is unknown, we assume the invariant not to hold and we first have to restore it, starting with status (UNKNOWN, UNKNOWN) (Fig. 2.2).

In automated applications the system has to maintain the invariant, i.e. restore the invariant after it has been disturbed. As the disturbances are described in the rules mentioned above, the system can determine when a restore operation has to be invoked. It does this by intercepting the operations the application performs on microscope settings and position controls. As these operations in the current architecture are already executed by the microscope system on request of the application, adding interception is easy and straightforward. The knowledge that the system has about the effect of the disturbance, i.e. state (CLOSE, OK) or (UNKNOWN, UNKNOWN), is used to determine which restore operation to use. Of course the restore operation used for (UNKNOWN, UNKNOWN) is always usable. However, it turns out that a restore operation that uses the knowledge that only the focus has to be restored (when the system is close to focus) is much faster. See also Sects. 3.1 and 6.3 where the two restore operations are described.

Note that relaxation effects and external influences might also cause the system to drift away from the optimal image controls and hence disturb the invariant. However, within the range of magnifications used in our experiments (see Sect. 3.1) and the time frames of the type of automated applications we are interested in, it shows that focus and astigmatism drift is negligible. A time-out value on the time the invariant holds might possibly be required in other applications, but was not relevant for this project. Position drift, however, is another relevant disturbance (see also Sect. 7.3). The correcting position drift might cause a disturbance of the invariant, as during executing this correction the rules on invariant disturbance also apply!

Sensor and Payload Data

From the above it is clear that to be able to acquire an image with suitable quality for the microscopy application we first have to acquire a number of images to restore the invariant. We will refer to the image we want to acquire for the application as payload data (image), and images to restore the invariant as sensor data (image).

Of course it would be very nice if we had independent sensors to measure the image quality directly. This would allow a control loop to be built so that the system is always ready to acquire payload images. However, such a sensor is not available.[2] In current electron microscopy the only sensor available with sufficient sensitivity is the instrument itself. Consequently we have to use the instrument both for acquiring payload data and sensor data. As well as being required for restoring the invariant, sensor data is also needed to allow for drift correction when acquiring payload data over a long period of time. This is the case in EDX, X-Ray data collection, or tomography.

Sensor data serves a different purpose than payload data and therefore has different requirements with respect to properties such as image content, number of pixels, and acquisition time. For instance, in order to restore the invariant it is not necessary to acquire sensor images of the same size as the payload image. We could use a smaller size, which is faster in acquisition time and speeds up the overall process. However, there should be sufficient information in the image for robust operation of the invariant restore operation. Consequently, selecting an area from the full image to be used for this purpose might not be straightforward. Furthermore, changing the settings and position controls in order to switch between sensor data and payload data should not trigger any change in the invariant. This means the freedom to define sensor data acquisition is bound by the intended acquisition of payload data.

Sensor and payload data acquisitions sometimes have to be interleaved. The interleaving patterns are determined by the simultaneous execution of multiple invariant restoring operations. Some sensor data acquisitions, such as for optimizing focus or absence of astigmatism, have to be finished before the payload acquisition. Other corrections, such as drift compensation, need sensor data acquisitions inherently interleaved with the payload data acquisition. One example is the repeated switching to sensor data acquisition *during* payload acquisition: after a few scanned lines of the payload image, each time a small sensor image is acquired. This sensor image is used to estimate the drift. The influence of the interleaving pattern may play an important role when dynamics are concerned. Luckily most microscopy applications investigate static samples. The interleaving pattern is also important for the control system as the sensor images will not be available at a regular pace. Therefore the control system should be able to cope with this variability.

---

[2]   Although in Sect. 6.1 a sensor that is very helpful is introduced.

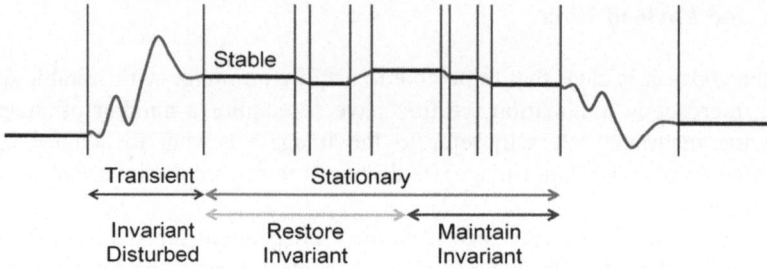

Fig. 2.3 The value of a key system parameter over time, showing the various phases

A Transient, Stationary or Stable System

Another characterisation of the system state is useful for our purposes. A change of system settings takes time as physical changes have to take place. For instance, a change in the lens current to re-establish focus is not instantaneous because of the lens dynamics. The time between applying the new current to the lens and the moment the lens is stable again is referred to as a transient. During transients all data acquired is considered to be invalid, for both payload and sensor data. This corresponds with our requirement that acquired data is only valid when the invariant holds.

Figure 2.3 shows the behaviour of a key system parameter, such as focus, over time. The large change is called transient, leading to a stationary state. In this state small changes are allowed to either restore or maintain the invariant, therefore stable phases are interchanged with short, unstable phases. Transients are intentionally applied to the system and usually explicit, whereas the small changes in the stationary phase are implicit and performed autonomously by the system. Note that more invariant restoring operations may occur in a stationary phase.

Using the above-defined concepts, the system controls its behaviour according to the model of Fig. 2.2. However, we need signals from the system to determine the beginning and end of transients and states. Dedicated sensors (e.g. magnetic field sensors) may provide input to models that describe key system parameters from which signalling could be derived, providing a signal at the start and end of a transient. These signals can contribute to driving the state model.

The combination of design concepts described in this section will help in creating a new systems architecture for an automated electron microscope. The concepts have to be extended, refined and worked out in more detail to finally realise a predictable and well-behaved system. And more concepts will have to be developed. However, with these concepts we were able to run the application of automatic particle sizing, as described in Sect. 5.2, by implementing the concepts on the experimentation platform in the Concept Car (Sect. 3.1).

Roughly the system behaves as follows:

The system maintains the invariant state, initially (UNKNOWN, UNKNOWN), by intercepting all microscope setting commands and all positioning commands invoked by the application. When the application invokes an acquire image command, the system checks the invariant state. Then, if required, it invokes the procedure (see Sect. 6.3) to restore the invariant, unnoticeable to the application. This procedure acquires sensor images as desired. After restoring the invariant the system acquires the requested payload data and returns this data to the application. During payload acquisition the system corrects for drift by interleaving payload data and sensor data acquisitions.[3] After providing the payload to the application, the application proceeds, performing the desired analysis and then invoking new commands intercepted by the system. This possibly results in a disturbance of the invariant again, etc.

Independent of the procedure above, all invocations of commands that change the lens settings, be it from the application or internal in the system, e.g. to correct for focus, wait until the system is stable before proceeding to a subsequent acquisition of sensor or payload data. The implementation of this may be simple, based on a worst case time out, or very advanced, using a feedback signal from the control unit. However, the concept remains the same. Note that in some use cases images may need to be acquired at high frequency, like in full interactive control by the operator. At these times the system should provide direct visual feedback to the operator. Also, payload acquisition will run continuously, overriding the aforementioned under the hood procedures. This provides backwards compatibility with respect to user behaviour.

Acknowledgments The author likes to acknowledge Bernard van Vlimmeren and Hugo van Leeuwen of FEI for their valuable discussions.

Reference

1. G. Muller, *Systems Architecting-A Business Perspective* (CRC Press, Boca Raton, 2012)

---

[3] This is only possible in scanning acquisition mode, not in camera acquisition mode.

Chapter 3
Feasibility Prototyping

Abstract A powerful method to investigate new architectural concepts is to create feasibility prototypes. These are known by a wide variety of names in industry: *concept car*, *development mule*, *alpha demo tool*, etc. Depending on their implementation, these allow various aspects of a new design to be investigated and evaluated as if the function or property already exists in the product. This chapter discusses the structure, advantages and challenges of three prototypes of a transmission electron microscope: (1) the feasibility prototype, which includes an approach to step-wise develop advanced system functionality in a project context, (2) the proxy device, an add-on system that minimally interferes with a current electron microscope while allowing many significant experiments on image-based global control, and (3) the microscope simulator, which allows investigation of local and global control strategies

Keywords Feasibility prototyping · Electron microscope · Experimentation platform · Data acquisition system · Scanning module · Proxy pattern · System simulation · Control architecture · System dynamics

3.1 Concept Car

Richard Doornbos

Embedded Systems Institute, The Netherlands

R. Doornbos and S. van Loo (eds.), *From scientific instrument to industrial machine*,
SpringerBriefs in Electrical and Computer Engineering,
DOI: 10.1007/978-94-007-4147-8_3, © The Author(s) 2012

Introduction

Generally, when research projects begin, the team often set out on a very complex mission. The intention of management or the project owners is that by setting the initial goals and expectations high, the project will accomplish more and the results will be more significant. Combined with a small set of somewhat vague and sometimes unfunded requirements, this usually results in little progress in the first stages of the project and almost no significant industrial results.

The current TEM automation project also started this way: a complex microscopy application was chosen as the leading case. This was a huge challenge given that an electron microscope is very hard to understand and operate for non-experts. Furthermore, there is no easy access to the internals of the electron microscope system, which is needed for performing experiments on system aspects. The system was not designed that way, and human and system safety should not be compromised by any system modifications. Finally, as the microscope is a rather expensive system (for purchase, as well as for use and maintenance), its availability for experimentation was limited. In summary, there was a high entry barrier for performing multi-disciplinary research in this field.

When we now consider the process aspects of running a research project in this field in more detail, we can better characterise the issues. Firstly, having a disparate group of academic researchers from different disciplines and research directions, with an inclination to theoretical work instead of practical application, makes true cooperation an issue. Also, the often overlooked factor of working at different locations can have adverse effects. Secondly, the industrial partner as a project owner is often a very critical customer as its employees are the top-experts in the field. This may be combined with the, generally shorter term, focus and interest of running a business creating commercial products.

So it is understandable that getting the project team into a cooperating mode with the industrial partner is hard. Furthermore, in general, academics and engineers are trained to be 'discipline thinkers' and not 'system thinkers'. However, it is the latter we need for solving a multi-disciplinary system problem (see Chap. 2).

We can conclude that this calls for an expedition-type of investigation approach. When we look at the process aspects of this investigation we can clearly distinguish four goals:

1. Convince the customer. Building and presenting prototypes to the customer and project owners helps to show the value of the research results and enables the project to alter the required direction of the work.
2. Stimulate research work. By having a leading case that focuses and guides efforts and sets boundaries, it is easier to achieve more cooperation between project members, which in turn is expected to yield more innovative or better applicable research results.
3. Learn (more) by doing. It is hard to understand essential system issues from paper alone. It is easier to obtain requirements, system settings, application

conditions, system conflicts and consequences, etc. when actually working with a realistic prototype.

4. Provide experiment support. It is important to lower the barrier for creative experimentation by providing easy access to the microscope's functions and data.

When we look at the technical aspects we can clearly distinguish three goals:

1. Characterise the system. First, we want to measure how the system behaves in typical circumstances. This helps us understand the issues that should be investigated further and what type of solutions should be developed.
2. Investigate feasibility of new concepts. The feasibility prototype is mainly built for investigating new concepts. These can be reasoning concepts, like sensor and payload images, or architectural issues, like how and where to add control and correction mechanisms. They can also be application level aspects, like workflow definition, decomposition of required advanced functions, and experimental design (see Architectural Patterns).
3. Prove advantageous cooperation between components. This is the central idea in systems thinking: the whole is more than the sum of its parts. We aim to show the possibilities that arise when advanced functions are added to the system, and especially when these functions depend upon each other.

A very clear need arises from these goals. To be able to perform characterisation of microscope behaviour (which can be used to validate models) and to create proof-of-concept prototypes of research results, we need an instrumented microscope system. An instrumented system allows changes to settings of the major components at the level required. It also allows intervention in certain control flows or interception of data signals.

Unfortunately, a fully instrumented system was not available.[1] Only access to standard products used in the R&D and engineering department, for testing the latest software upgrades or recently developed components, could be provided. Changing parts of these systems was not a feasible option: implementing changes to these (permanently) running test systems would mean investing a large amount of development effort, significant involvement in the engineering process, interfering with scheduled tests, and accepting the financial risks involved (after all, things can go wrong in experiments).

Approach

The 'Concept Car' approach aims to solve the above problems by performing step-wise integration of research results on the machine. The step-wise approach separates the problems and concerns in a sensible way, and this simplifies matters

---

[1] In the high tech industry this is very often the case.

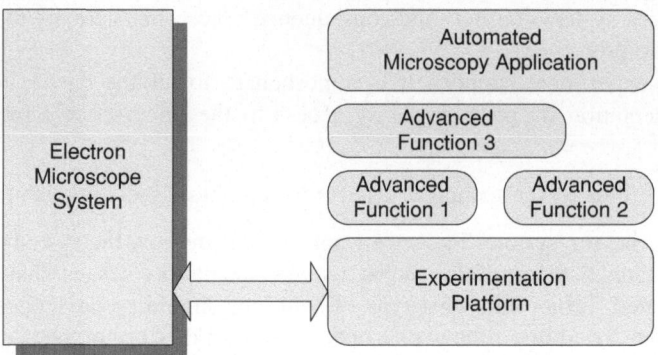

Fig. 3.1 System structure of the Concept Car prototype with a standard electron microscope (product) on the left side

considerably. Furthermore, step-wise integration is performed with organic growth in mind, so eventually an advanced functioning system will have been created.

We will discuss the Concept Car approach on three levels (Fig. 3.1):

- Microscopy applications (e.g. automatic sizing)
- Generic advanced functions (e.g. automatic correction)
- Experimentation platform

First of all we define a set of microscopy applications as intermediate steps, called *Concept Cars*, with each step becoming more complex, and all leading towards the catalyst microscopy application. These intermediate steps address both application goals and more technical goals. The application part determines the microscope settings and needed functions, quantifies the requirements and contains application research (Fig. 3.1).

The second part—containing most contributions from the academic partners–provides the technical support in terms of advanced functionality for the applications. These contributions are in the form of application elements, correction algorithms and control algorithms. All are for mechanical stabilization, automatic focus finding and automatic optical optimization. The definition of these generic advanced functions depends on the selected microscopy applications, the output of the characterisation experiments, and the knowledge and experience of the industrial experts in the field, i.e. the FEI engineers.

Thirdly, we develop an experimentation platform (*Expla*) by realizing a very low invasive instrumentation mechanism that provides flexible and easy access to the standard system, and does not require changes. This instrumentation platform allows us to characterise the microscope behaviour, as well as building more complex microscopy applications.

Concept Car definition requires considerable insight into the overall problem space, which covers the microscope as a system, microscopy applications, as well as the contributing technical components and their correct application.

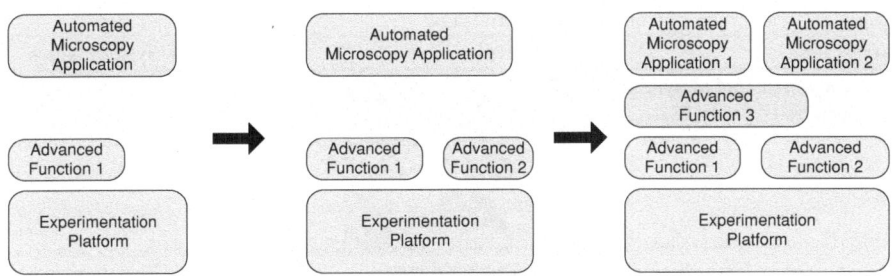

Fig. 3.2 Step-wise approach leading to Concept Cars

Figure 3.2 depicts the step-wise approach, in which the development and integration activities lead to Concept Cars with increasing functionality.

Microscopy applications

The ultimate case that we selected as a leading microscopy application was the analysis of catalytic particles. This leading use case is too complex to address directly: it uses very high magnification, multiple complex microscope modes, and requires sophisticated image processing technologies. However, as an end goal, it was both realistic and provided a challenge for innovative thinking. We defined several intermediate steps, for which the microscopy applications and the addressed system issues are shown in Table 3.1. The associated platform and advanced functions are not indicated.

Concept Car 3—Size Distribution Measurements

This Concept Car takes up several challenges, at application as well as systems level. The microscopy application's task is determining the size distribution of a mixture of cement or gold nanoparticles. Determining the size from electron microscopic images is a challenge on its own. Just finding the particles in the image is difficult, as is the inverse problem of reconstructing the particle size from the image formation process.

The size distribution within the specimen is deliberately chosen so that the particles cannot be measured at a single magnification setting, leading to magnification changes during the measurement procedure. Furthermore, we want to measure a large amount of particles to obtain an accurate size distribution. This calls for measuring images at many locations on the specimen. We know from performed characterisation experiments on the microscope that both changing magnifications and changing locations using the stage will lead to system disturbances and changes of operating point. These require correction or re-optimisation.

Optimizing the overall throughput is a challenge, affecting all parts of the application and system. We addressed this issue by using an adaptive sizing

Table 3.1 Definition of Concept Car applications

Application	Specimen	System issues
Size measurements of synthetic particles (Agar S128)		Automation, throughput
Size & shape measurements of $MgFe_2O_4/ZrB_2$		Real samples, accuracy
Size measurements of a mixture of cement, or synthetic gold nanoparticles		Scalability, automated corrections
Catalyst case (difficult): shape measurements & chemical analysis of Pt/TiO_2		Long duration, electric/ mechanical stability
Catalyst case (very difficult): shape measurements & chemical analysis of Ni-La/ZrO_2		Complex system modes, particle recognition

algorithm, which achieves an optimum throughput by adapting the measurement settings *during* the procedure (see Sect. 5.2).

Application Details

Determining the size distribution of a mixture of, for example, 5, 30 and 90 nm nanoparticles consists of several tasks. In short, this is the terrain of experimental design, a special field in electron microscopy concerned with determining the

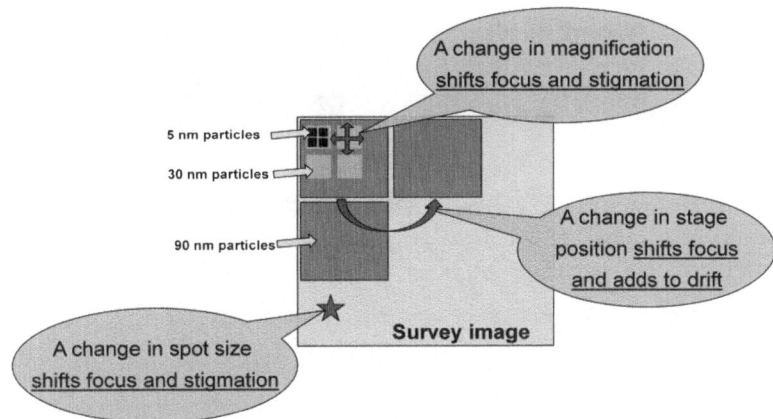

Fig. 3.3 Significant disturbances are introduced by changes to the parameter settings of the electron microscope

optimum measurement procedures and workflow given the required information needs. In this case the measurement procedure consists of setting the microscope to optimum measurement settings, capturing images, finding particles in the image, determining accurately the size of each particle, and keeping track of the locations and results. While designing this workflow it is useful to develop new organizing elements: for example we defined 'units' to describe certain phases of the workflow.

Generic Advanced Functions

The required set of generic advanced functions was derived from characterisation experiments. An overview of the outcome of these experiments is shown in Fig. 3.3. From these results it is clear that several on-line corrections of system parameters are required for the current application (see Fig. 2.2 of Sect. 2.2). We decided to create a fast focus correction method (see Sect. 6.3) to be used after a small stage move; an extensive focus and stigmation optimisation (see Sect. 6.3) to be used after significant changes of parameters like magnification and (beam) spot size; and a mechanism to counter image drift (see Sect. 7.3).

Experimentation Platform

The basic requirements for this platform are: easy access to all microscope functions, programmability, quick set-up, no relevant intrusion of the electron microscope, system-safety, and little effort to build.

The options to create such a system are limited. Therefore we rely on thin interfaces, reuse of existing systems, and commonly available resources.

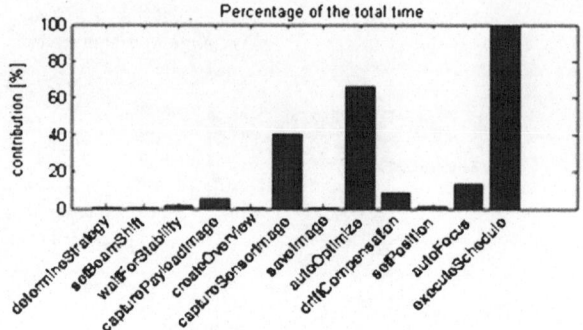

Fig. 3.4 An example of a statistical analysis to identify activities that create bottlenecks

To mention a few examples: microscope access was made using the TemScripting interface, a DCOM-based interface that can be used from the Ethernet network level. We implemented it in such a way that the user can call these microscope objects using Java. We also provided the option to access the microscope from scripts in Matlab, the default computation tool of many engineering and academic researchers. This entails allowing the user to access Matlab libraries for advanced computation, e.g. the image processing toolbox. Finally, we can mention some additional strong features of Expla as zero intrusion, enabling remote operation and operating system independent.

Technical Results

The technical goals of acquiring sufficient requirements and system information while developing the system, enabling feasibility tests of new concepts, and showing the advantages of cooperation between components were fully achieved. The system also allowed a 'qualitative' throughput and workflow analysis by visualizing the actual procedure execution, combined with statistical activity overviews. These clearly pointed out which elements were bottlenecks (see Fig. 3.4). By reasoning about translation to a more realistic implementation we could get a good understanding of the major factors influencing the system throughput.

Inherent in precision critical systems are the system's reactions to environmental and internal disturbances—this is sufficiently addressable using the current approach. The microscopy application repeatedly invokes these unwanted reactions by the system. This means we can investigate sound approaches for re-establishing the right conditions for measurement, and for correction or re-optimisation algorithms.

Of course there are also some downsides in taking this approach, mostly due to the practical limitations inherent in the chosen set-up. First of all, the microscope is accessed via a network interface to the software layer that provides access to the hardware and its functionality and safeguards system safety. However, this adds extra delays and latencies in the execution of activities. For some experiments, such as correcting high frequency vibrations, the system is just too slow. And—due to the Expla 'add-on'—accurate temporal analysis was not possible (although estimations were!). Also, no direct mechanical and electrical hardware steering for control experiments was possible.

New Technical Insights

It is hard to summarise all the insights discovered, at application and system/ architectural level. Therefore we only mention the most important new insights.

At application-level most new insights were gathered in creating the workflow, mainly because the practical approach of proposing and testing led quickly to results.

At system-level many architectural aspects became tangible. First, creating such a prototype provides an overview of the system, which helps in the reasoning about the system, and anticipated systems. In a way the Concept Cars contributed to the recognition of new architectural patterns and solutions for the anticipated electron microscope system (see Sect. 2.2). Also, communication with the customer was strongly helped by displaying hard-to-imagine architectural concepts.

One eye opener was the observation that measurement throughput is seriously degraded by control and correction activities needed to keep the system in optimum state (see Fig. 3.4). This suggests adding strict time requirements on (future) correction and optimisation algorithms.

Another important new insight is that a very high throughput can be achieved by relying on mathematics-based measurement procedures. When using the right physical and mathematical models, sizing of particles can be made using only a few pixels per particle. This is in strong contrast with commonly used methods which use large overfill factors (see Sect. 5.2).

From all our development efforts it became clear that there is a great need for robustness on system level as well as on image analysis level, to achieve a truly predictable and reliable routine machine.

Process Results

Our observation is that by using the Concept Car approach more concreteness and focus is felt by the project members. Also, an increased level of cooperation and synergy emerged. We believe that the increased understanding of, and experience with, the microscope has led to innovations.

Furthermore the Concept Car prototype system provided valuable support in validating other, related, research results from some of the (academic) partners in the project.

The step-wise approach gave adaptability and flexibility during the development of the Concept Cars. For example, type of functions, level of needed accuracies, or time available could be adjusted to the constraints at that particular moment.

The Concept Cars were used as demonstrators in the communication with management and peers within FEI, as well as within the ESI network. These demonstrations were given regularly, typically at the end of each Concept Car integration activity.

There is considerable industry acceptance of the Concept Car results as it stimulated the recognition of the need for instrumented systems. It also provided a lot of inspiration at application and realization level.

We should be fair and show that there are also some less positive points about the Concept Car approach. For example, significant effort is required to create a working prototype, and there is a risk of focusing too narrowly on certain aspects, which might not be the most important ones. A well-known pitfall in this context is the so-called 'PR trap': spending too much time on polishing a prototype for demo purposes. We believe we did not fall into this trap this time.

Inherent in large and complex high tech systems is that the team is dependent on professional support and the availability of a dedicated system in industry. This means that the characterisation experiments are performed only on a single development test system (which is changed very regularly). Therefore conclusions about system performance of other systems cannot be made. Another observation about development test systems of this type (usually in R&D departments) is that they are hard to keep operational 100% of the time.

New Process Insights

Although there was ample opportunity to influence the direction of the Concept Car work, only limited discussion on technical issues took place with the project owners and key industrial contacts. This required leadership to make some decisions when needed, and therefore sometimes adverse decisions were taken, e.g. choosing for technology that was preferred by the researcher (Java and Matlab), instead of the platform used by the customer (C, C++). For a successful 'landing field' we may need a different decision, but in this case it was a balance between project progress and acceptance by the customer.

The step-wise aspect was particularly expected to generate willingness in *all* partners to contribute, but this proved unrealistic. Some partners were not fully addressed by this approach due to a mismatch between required and provided functionality, an awkward timing with respect to already running investigations, or the perceived lack of connection to 'real' scientific work. This holds in particular for partners working on component improvement, and especially on system and software modelling.

Challenges

Another important conclusion can be drawn from the Concept Car work. Although we tried to involve academics in industrial problems and practices, extensive validation in industrial practice and improving industrial applicability has a low priority for many academic researchers. The recognition of scientific work is mainly found in journal publications and conference contributions.

A second issue is the fact that creating such prototypes leans towards applied science and not many academics are trained in that direction. In addition, pragmatic researchers are required to do the job. This, coupled with the fact that sometimes Expla did not fit to the project or particular experimental goals, showed that a good mutual understanding is needed between partners.

Conclusion

Overall, we are convinced that the Concept Car approach worked well. All goals were met to a large degree:

- Show the generated synergy by the integration of scientific work: improved system behaviour has a cumulative effect
- Provide a realistic context for academic contributions
- Provide a platform for experimentation and learning
- Enable validation in industrial practice

As the industrial customer was impressed by the progress, transferring Expla may improve flexibility and speed for some investigative work made by researchers in the R&D and engineering departments in FEI.

Acknowledgments The author likes to acknowledge Rients de Groot of FEI for his efforts to keep the development microscope up-and-running.

3.2 Taking Control Using a Proxy

Nol Venema

Technolution BV, The Netherlands

Introduction

The ability to execute experiments on an operational microscope is one of the essential requirements for the success of the Condor project. Expla was designed to execute experiments on operational microscopes (see Sect. 3.1). Expla gives easy access to the scripting interface of the microscope application on the Host PC. The scripting interface provides basic control of the electron microscope such as moving the stage, changing magnification and capturing images. This high-level interface is somewhat slow. However, some experiments require detailed control over the scanning and acquisition process. For example, an on-the-fly autofocus algorithm that refocuses after every 10 payload images by quickly scanning a

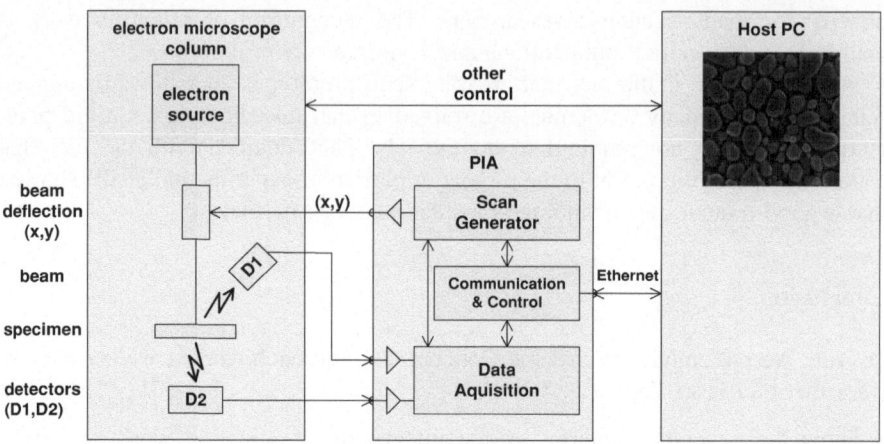

Fig. 3.5 The PIA in its context

series of 20 smaller sensor images while changing the focus settings. Therefore our problem is that the microscope lacks the instrumentation needed for elaborated experiments concerning low level control of the scanning acquisition sub-system.

Context

A PIA (Patterning Imaging Acquisition) is the scanning device in FEI electron microscopes. Figure 3.5 shows the PIA in the context of the microscope. The basic scanning process consists of two main functions: generating the deflection signals for deflecting the electron beam and sampling the detector signal for every pixel of the image. The assembled images are forwarded to the Host PC system, which controls all microscope components and provides the operator's user interface.

Although scanning rectangular images is the most common task for the PIA, it is certainly not limited to this feature. The PIA contains all features needed for delivering high quality images, e.g. excellent real-time characteristics and flexible scanning control strategies.

Solution Concepts

Typically the Host PC is needed to do all kinds of preparations, like finding a suitable area on the specimen and selecting basic settings like magnification and focus. Only after all basic settings are done (e.g. by the operator) do we want to take control of the PIA, execute the experiment and return PIA control back to the Host PC.

Taking over control of the PIA in an operational microscope is not an easy task. The following three approaches could result in viable solutions to our problem (Fig. 3.6):

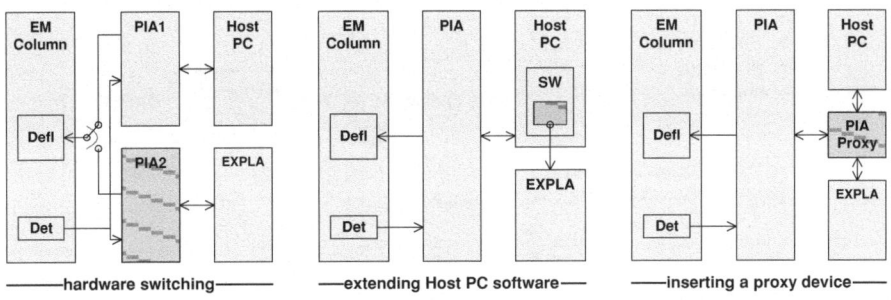

Fig. 3.6 Three alternative solutions

1. **Hardware switching**: connecting a second PIA to the column and to Expla
2. **Extending Host software**: extending the Host PC software and connecting it to Expla
3. **Proxy device**: inserting a proxy device between Host PC and PIA and connecting it to Expla

The 'hardware switching' solution switches control signals between two PIA systems. During the experiment, the PIA2 connected to Expla should be switched inline. Major drawbacks include the large engineering effort needed to enable switching, and the fact that Expla needs to know the state of PIA1 to bring PIA2 into a compatible state.

The 'extending Host software' solution integrates software for the experiment in the software stack of the Host PC system. Unfortunately software integration in such a platform is complex, version dependent, requires significant effort and could compromise the stability of the Host system because of potential bugs in the software for the experiments. It would also require changing the PIA driver for external access to Expla (Fig. 3.6).

The 'proxy device' solution inserts a PIA-Proxy system at network level between the Host PC and PIA (an application of the 'man in the middle' architectural pattern). The PIA-Proxy software provides a PIA interface to the Host PC and passes all messages to the PIA in its transparent mode. Expla has access to the control interface of the PIA-Proxy and can switch itself to control the PIA. In this mode, the Host is effectively connected to a software simulator of the PIA and is unaware that it is no longer in control.

Comparing solutions, the last seems the most attractive option because the installation procedure is easy, only involving connecting some Ethernet cables. No further dependency exists and no changes are needed on either the microscope's hardware or software.

On the other hand, for the proxy solution to succeed, the following requirements should be met:

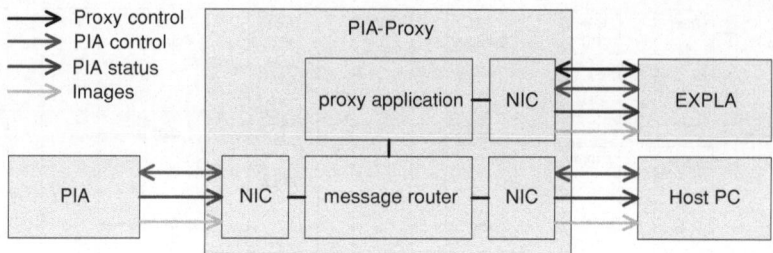

Fig. 3.7 PIA Proxy configuration

- The network protocol specifications should be available and application building requires knowledgeable engineers.
- The protocol software (e.g. the microscope application on the Host PC) must cope with a slightly different timing behaviour of messages.
- The proxy-system must deliver certain throughput requirements (e.g. image rates and message sizes).

PIA-Proxy in More Detail

In the proxy solution, the well-known *proxy* architectural pattern is used between two communicating parties. For these parties, a proxy offers a similar interface as the device it hides. Because all communication to a device passes through the proxy first, it basically gives the proxy absolute control over the device.

Figure 3.7 depicts the complete set-up's structure and data and control streams. Note that Expla will also be connected to the PIA-Proxy to enable control of the proxy and the PIA, as well as for receiving intercepted images. The PIA-Proxy consists of two main components: the message router and the proxy application. The message router is connected to both network interfaces of the PIA and the Host. It receives and transmits messages on these interfaces and allows the proxy application to intercept and inject messages. Configured for a particular experiment, the proxy application typically intercepts specific messages for further handling, like passing, modifying, blocking and sniffing.

Figure 3.8 shows the three standard message exchange types between PIA and Host PC in an UML sequence diagram. Command and response messages are used to configure and control the PIA. Status messages are sent by the PIA to inform the Host PC about its status. Image messages are sent in huge bursts for transferring the images from PIA to Host PC.

A simple use case is 'sniffing': just sniffing images on behalf of Expla, shown in Fig. 3.9. All messages are passed transparently except for the image messages, which are intercepted, copied and forwarded to Expla.

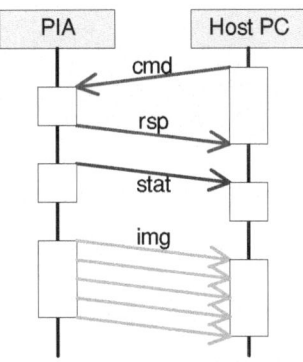

Fig. 3.8 Message exchange types between PIA and Host

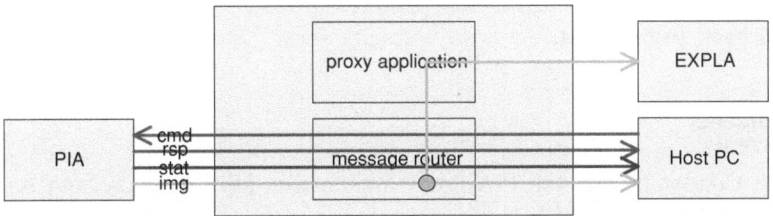

Fig. 3.9 Image sniffing use case

A more interesting use case is 'single sensor image'. In this scenario, Expla temporarily takes over control, executes a particular scan job and switches back to normal. It is very important that the Host does not notice any of the activities, because we do not want the protocol handling software to get in an error state and halt. Figure 3.10 shows this use case at the moment that Expla has taken over control, while Fig. 3.9 represents the situation before and after the experiment.

The proxy application contains two main components: experiment logic and a stub. The experiment logic is responsible for mode switching (between normal and experiment mode) and activities during the experiment. During the experiment it typically downloads a scan job and executes it once. The image from this scan is sent to Expla only. After the special scan, the proxy switches back to its normal mode and restarts the original scan job. The Host receives fresh images and fully controls the PIA again.

During the experiment the stub is active to keep the Host functioning correctly. The stub's main task is to simulate a PIA in a state that the Host expects. Therefore it sends the Host fake status messages and image messages (which might be the last image sent before the experiment started). Although command

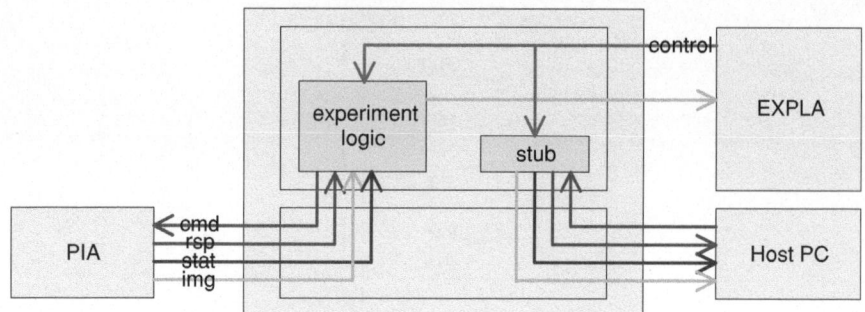

Fig. 3.10 Advanced experiment use case

messages are not expected in this situation, commands are all handled by the stub. After the experiment the stub restores the state of the PIA and switches control back to the Host.

Experiments

For the Condor project the PIA-Proxy was successfully used on both scanning electron microscope (SEM) and scanning transmission electron microscope (STEM) systems of FEI. The following experiments were designed using the PIA-Proxy solution:

- **Fast scanning**. An experiment with insertion of a scan job that scans a small area of the payload scan area with a high frame rate (e.g. 100 Hz on a STEM, 10 kHz on a SEM) for a specified time. High frequency disturbances can be easily visualised in this way, see Fig. 3.11.
- **Sensor scanning**. An experiment with repeated insertion of a single sensor scan, after every 5 payload scans for the Host. The sensor scan data can e.g. be used for drift correction.
- **Time domain sub-sampling**. An experiment with the insertion of a special scan job that samples a small area of the payload scan area. For every pixel, a series of samples at a higher sample rate (e.g. 8 times)—thus using a shorter integration time—is acquired. This can be used to analyse the counting statistics.
- **Drift measurement**. For this experiment the original scan job is extended by embedding additional reference scans. This means that a small reference area is scanned for every 10 lines in the payload scan. The resulting sensor image will be sent to the drift correction module in Expla, which determines the drift vector and modifies the acquisition settings or corrects the original image (according to a drift model). The Host is unaware of this and only receives the image data it expects (the reference scan data is stripped out by the PIA-Proxy).

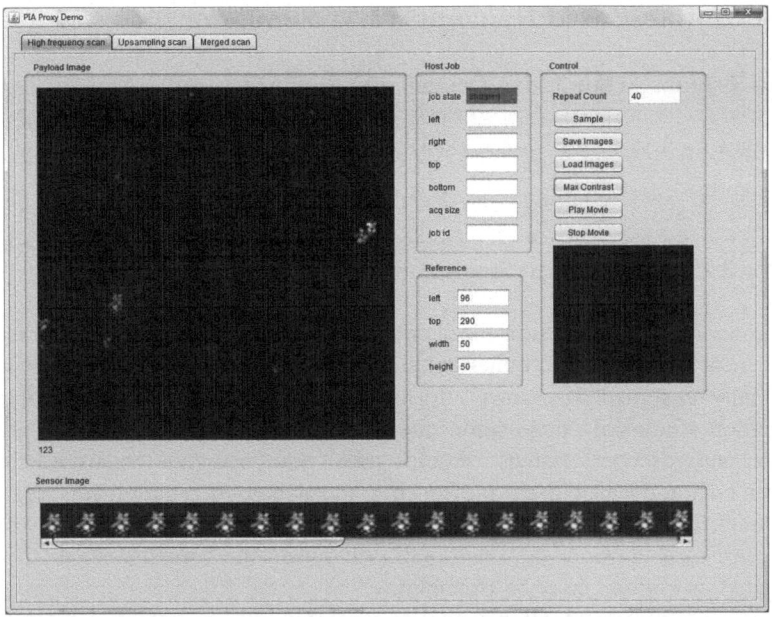

Fig. 3.11 Screen shot of the 'fast scanning' experiment on a STEM, showing at the bottom the images obtained at 100 frames per second

Conclusion

We created a viable solution to the problem of accessing and controlling a scanning acquisition system in the electron microscopes of FEI for performing system experiments. The PIA-Proxy provides a non-intrusive method (sniffing) for obtaining images that can be used for on-line correction and control. It also offers a slightly intrusive method (with respect to temporal behaviour) that allows full PIA access and control.

The experiments carried out clearly show that designing and performing experiments using the heart of a scanning microscope was fairly easy, required little effort and could be done without FEI support. The results also indicate that having an instrumented electron microscope is essential for developing system innovations quickly and cost-effectively.

The proxy solution might also be interesting in other cases in which the system under investigation is complex, closed or otherwise inflexible, and one needs essentially an instrumented system, i.e. direct access to sub-systems, control channels and data streams.

Acknowledgments The author likes to acknowledge Gert Jan de Vos of FEI for his support.

3.3 A Matlab-Based Dynamic TEM Simulator

Arturo Tejada, Arjan J. den Dekker and Paul MJ van den Hof

Delft Center for Systems and Control, Delft University of Technology, The Netherlands

The Need for a Dynamical Simulator

Next generation transmission electron microscopes (TEMs) will become automated measurement tools rather than image generating devices. They will be
specifically designed to extract information from specimens, like particle size
distribution, chemical composition and structural information. Therefore, future
electron microscope designs should take into account electro-mechanical
requirements and particularly the sensing and actuation requirements needed to
implement the feedback or feed forward control loops necessary for automation.
They must also consider the coordination between the constituent parts, so that
automated procedures may be executed.

A software-based simulator would be a helpful and cost-effective tool to study
different design alternatives, providing insight into what to design and how to
design it. It needs to take into account the dynamical properties of the components
of interest and allow easy testing of different closed-loop control ideas. Such a
simulator should be capable of recording the temporal responses of components.
Also, in the ideal case, it will simulate the video stream that the microscope would
generate under the observed component responses. Eventually, the simulator
should provide insight into the way current components and processes can be
re-utilised or reconfigured to aid in the automation process.

Implementing such a simulator requires an understanding of how the different
internal components interact when setting particular variables of interest. For
instance, the amount of defocus is set by the current applied to the objective lens,
perturbed by the position of the specimen in space which, in turn, is determined by
the specimen holder and measured from images [1, 2]. In addition, it requires an
understanding of the components' temporal (i.e. dynamic) properties, e.g. their
reaction speed, and knowledge of their temporal sequencing, i.e. the order/timing
of component operations.

Implementing such a simulator for a TEM was started using Matlab's Simulink
tools. Its architecture and capabilities are summarised next.

Simulator Architecture and Capabilities

The first step towards developing the simulator was to provide a model of the
functionalities of current TEM components and their relations from the point of
view of a control engineer. Figure 3.12 shows an example of such a model.

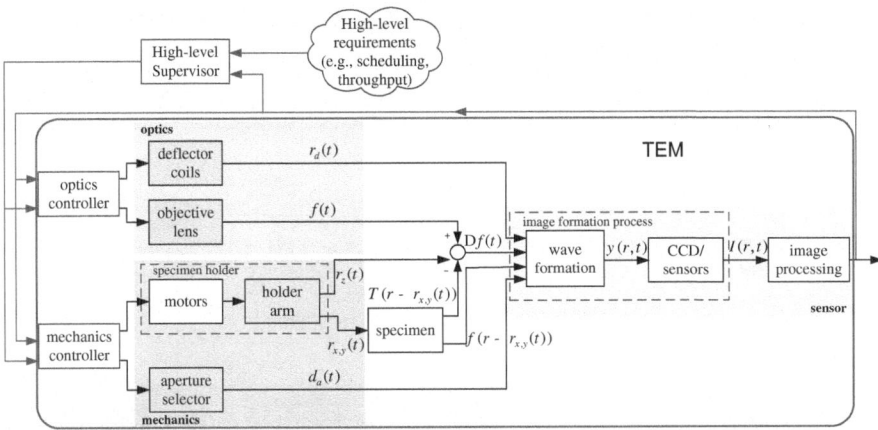

Fig. 3.12 Functionalities of current TEM components from a control engineering perspective

The boxes indicate components or processes whose dynamical models must be identified from first physical principles or through experiments [1]. For some of these boxes the models are known, or have been identified (see Sects. 7.2 and 7.3).

Note that as long as these models are not yet available, for simulation purposes and testing the unknown models can be replaced by a sensible default behaviour, e.g. a constant gain.

Additionally Fig. 3.12 shows a two-tier control structure. The lower tier contains an array of local controllers in charge of regulating individual components or processes. The higher tier contains a high-level supervisor in charge of temporal and sequential coordination of the different components. It is also responsible for trade-offs between accuracy and performance, etc. The main difference between the supervisor and the local controllers is that the former generally displays finite dynamics. That is, in general it cannot be described by a differential or a difference equation. Instead, its dynamics are described by an automaton (see [6] and the next subsection). Finally, note that Fig. 3.12 also shows processes that are not easily described by either differential equations or automatons. These include, for instance, the image formation process, which is a statistical, time-varying counting process [3], and the image processing algorithms needed to estimate parameters of interest from images.

The second step towards the simulator development is to perform a timing analysis. Note from the above discussion that some components operate continuously over time (e.g. the objective lens). Others operate only when they are triggered by an event (e.g. the image processing algorithm is executed only when an image is available). Operation of the first type of components is simulated by time-discretizing the differential equations that model their behaviour, using a small time step related to the component's dynamics, relevant from a system's perspective [4]. On the other hand, operations of event driven components are generated by the high-level supervisor whose associated automaton in turn triggers

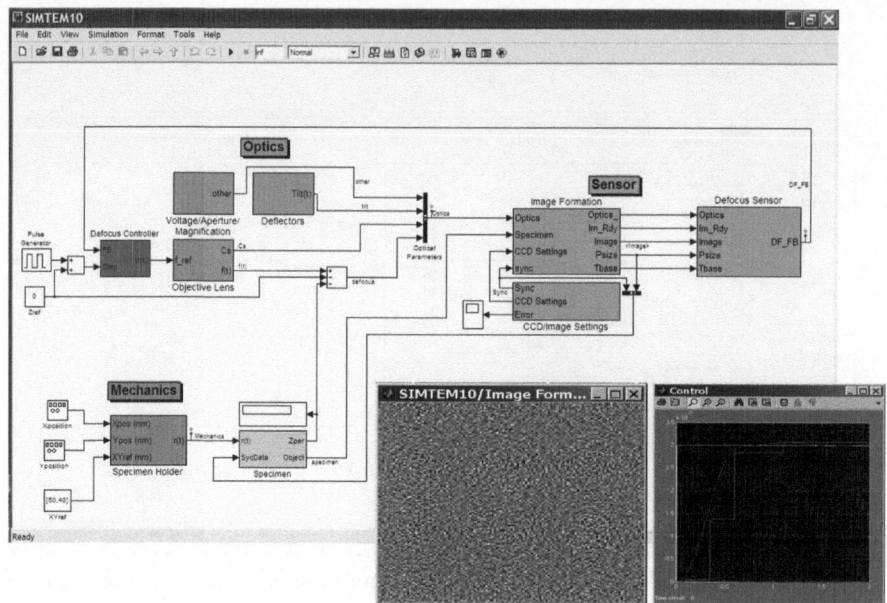

Fig. 3.13 Simulink implementation of the TEM dynamic simulator

the execution of function blocks. The function blocks themselves are used to execute complex Matlab code, such as image analysis, without affecting the simulation time (see [5] for details).

Once the dynamical models are discretised and the structure of the supervisor's automaton specified, the simulator can be directly implemented in Matlab's Simulink, as shown in Fig. 3.13. The simulated components have been interconnected following the architecture given in Fig. 3.12. This illustrates the simulation of current TEMs, allowing validation of the components and the current system behaviour. Adding the supervisor and advanced control algorithms (see Sects. 6.2, 6.3 and 7.3) allows the use of the simulator in forward engineering, to provide early validation of new concepts and implementations. The functionalities currently available in this simulator include [7]: live display of TEM internal signals (e.g. defocus, specimen position, etc.); 'live' image stream; online adjustment of magnification, defocus, beam tilt, specimen position, camera integration time, image rate, gun voltage, and camera pixel size; offline adjustment of image size; online defocus measurements; active coupling of the specimen's topography with the defocus level; and online defocus control.

Note that the temporal coordination between the continuously operating components and the event-driven component requires special attention. Although this coordination can be attained using ad-hoc methods and good engineering intuition, it is better to use systematic methods such as those in [6] to avoid developing a difficult-to-troubleshoot simulation. The next subsection provides some insight on these issues.

On the Coordination of Continuous and Event-Driven Dynamics

The events that trigger the response of event-driven components are issued by the high-level supervisor, which is in turn modelled by an automaton. The simplest automaton model is a finite state machine (FSM). Note that there is no timing information associated with an FSM. In practice, however, the input values do appear at particular times. If the time step used to time-discretise the differential equations is small enough (see previous subsection), it can be assumed with small approximation error that the input values appear at integer multiples of the time step. This allows one to treat an FSM as a very specific type of discrete-time system whose state changes every 'tick' of a clock (with a period equal to the time step) and to implement it in Simulink.[2]

Note that FSMs can also be used to aid in the temporal coordination of components, since they can divide the clock rate or provide counting mechanisms to trigger certain actions only after a certain number of clock periods (see e.g., [9]).

Finally, as stated in the previous subsection, some processes in a TEM cannot be easily modelled via differential equations, automata or FSMs. For instance, the image formation process involves counting electrons, each of which has an associated equation of motion that depends continuously on the microscope's time-varying optical parameters. The flight time of each electron is in the nanosecond order, which is about 1 million times shorter than the time scale of the main TEM components. Furthermore a large number of electrons is needed to form an image.

It is obvious that concurrently simulating the dynamics of the image formation process and those of the main TEM components is not a sensible way to go. The solution to this problem is to simulate these processes offline. For instance, to simulate the image formation processes the values of the optical parameters during a period of interest are provided to the off-line simulation, including the period of interest (e.g. 10 time steps). This data is then used by the function block to simulate and aggregate different image slices (one per time step), by approximating the optical parameters' variation with piecewise constant functions and using statistical image models (see [1, 3]). Note that, from the point of view of the simulator, the image slices are created offline in zero time (the Matlab function block consumes zero simulation time). Thus, the total image simulation time is equal to the period of interest (10 time steps in this example). In case additional delays are required (e.g. to simulate the CCD camera read-out time), they can be easily added to the simulator.

---

[2] Details on how to transform a direct graph into a discrete-time LTI system in Simulink can be found in [8].

Conclusions

Although the availability of a TEM simulator is very helpful, its realisation and validation turned out to be much more complex and required more effort than expected. Many of the necessary models, of both the current components as well as the image formation processes, were not readily available, or only available as theoretical models that were not tuned to the specific electron microscope involved. However, the overall modelling, to the level that implementation of the simulator became possible, created helpful insights that would have been difficult to obtain otherwise.

References

1. A. Tejada, A.J. Den Dekker, W. Van Den Broek, Introducing measure-by-wire, the systematic use of systems and control theory in transmission electron microscopy. Ultramicroscopy **111**, 1581–1591 (2011)
2. A. Tejada, W. Van Den Broek, S. Van Der Hoeven, A.J. Den Dekker, Towards STEM control: modelling framework and development of a sensor for defocus control, in *Proceedings of 48th IEEE Conference on Decision and Control*, (2009) pp. 8310–8315
3. A. Tejada, A.J. Den Dekker, The role of Poisson's binomial distribution in the analysis of TEM images. Ultramicroscopy **111**, 1553–1556 (2011)
4. G.F. Franklin, D.J. Powell, M.L. Workman, *Digital Control of Dynamical Systems*, 3rd edn. (Addisson Wesley, Menlo Park, 1998)
5. Mathworks, Using the matlab function block (2011), http://www.mathworks.nl/help/toolbox/simulink/ug/f6-6010.html
6. C.G. Cassandras, S. Lafortune, *Introduction to Discrete Event Systems*, 2nd edn. (Springer, New York, 2011)
7. A. Tejada, Measure by wire (2011), http://www.tejadaruiz.net/MBW/html
8. A. Tejada, Stability analysis of Markov jump linear systems with Markov inputs, Dissertation, Old Dominion University, Virginia, (2006)
9. R.J. Tocci, N. Widmer, G. Moss, *Digital Systems: Principles and Applications*, 11th edn. (Prentice Hall, NJ, 2011)
10. A. Tejada, J.R. Chávez-Fuentes, P. Vos, Stability and performance analysis of dual-rate systems with random output rate via markov jump linear system theory, in *Proceedings of the 50th IEEE Conference on Decision and Control*, Orlando, Florida, pp. 2895–2900

Chapter 4
Software Architecture

Abstract Specific models of the software architecture are required to predict the behaviour of a system. Unfortunately, only a limited number of accurate models are available, and no generally accepted modelling paradigms exist for many of the relevant aspects (especially for run-time execution and threading). One of the aspects considered problematic in the electron microscope system investigated was the large amount of threads created at run-time. The system was reverse engineered to model the thread usage, with the goal to reduce this amount significantly. This chapter describes the development of a modelling approach for this purpose, starting from the well-known viewpoint approach. Furthermore, it describes the application of this method to the electron microscope and a software system for validating the suitability of the approach.

Keywords Software architecture · Modelling · Concurrency · Parallelism · Architectural viewpoint · Thread pool · Performance bottleneck

4.1 Thread Performance Modelling

Naeem Muhammad and Yolande Berbers

DistriNet, University of Leuven, Belgium

Introduction

Developing high quality software is hard. Software Quality Attributes have been introduced in the field of software engineering to enable measuring the fitness and

R. Doornbos and S. van Loo (eds.), *From scientific instrument to industrial machine*,
SpringerBriefs in Electrical and Computer Engineering,
DOI: 10.1007/978-94-007-4147-8_4, © The Author(s) 2012

suitability of a software product. Functional quality reflects how well the software system complies with, or conforms to, a given design, based on functional requirements or specifications. Structural quality refers to how the software system meets non-functional requirements that support the delivery of the functional requirements. Typical examples of important non-functional requirements at runtime are performance, security, availability and interoperability.

In this section we are interested in performance as a non-functional property. It can be described as the system's response time, utilisation and throughput behaviour. More specifically, we are interested in the performance aspects that are related to the use of threads, the software entities consuming the available CPU cycles.

Performance is often ignored until there is a problem. This reactive approach has serious drawbacks as performance problems are frequently introduced early in the design. And, as design issues cannot always be fixed through tuning or more efficient coding, fixing architectural or design issues made later in the cycle is often difficult and therefore very expensive. However, as this is general practice in many projects, it is important to support it with adequate approaches. These help the system engineer to tackle the problem and understand the system's behaviour through 'reverse' engineering.

A software architecture description covers the fundamental organisation of a system embodied in its components, their relationships to each other and to the environment, and the principles guiding its design and evolution. Used to understand a complex system, it guides the realisation of the system in all phases.

An architecture is usually represented by a set of models that together provide a coherent description of the system. A single, comprehensive model is often too complex to be understood and communicated in its most detailed form, showing all the relationships between the various business and technical components. As with a building's architecture, it is usually necessary to develop multiple views of an information system's architecture. The different views enable the architecture to be communicated to, and understood by, the different stakeholders in the system [1].

For example, just as a building architect might create wiring diagrams, floor plans and elevations to describe different facets of a building to its different stakeholders (electricians, owners, planning officials), so an IT architect might create physical and security views of an IT system for the stakeholders who have concerns related to these aspects.

The system's 'concurrency structure' contains a mapping of functional elements to concurrency units. It clearly identifies the parts of the system that can execute concurrently, and how they are coordinated and controlled. The concurrency structure is an important element in understanding performance issues related to the use of threads.

Problem Statement

The software of an application like an electron microscope is composed of a set of programs, running on one or more CPUs. A thread allows a program to perform more than one task at a time. Depending on the type of program, multithreading can significantly improve the performance of a program. For example, programs that greatly benefit from multithreading are those that spend a great deal of time waiting for outside events, a typical feature of control software.

However, having many threads can also negatively influence the performance of the system [2]. If there are many short-lived threads that only deliver a small task, the system might spend the bulk of its time creating and destroying threads rather than simply doing its intended work. Creating and destroying threads are expensive operations that do not add to a system's functionality. Another issue is that the more threads active in a system, the more time wasted on switching between them. For these reasons, analysing the behaviour of threads in a software system is important to gain insight in this problem domain.

A possible solution is the use of a thread pool. Thread pooling is a multithreading technique in which threads are reused from an existing pool, instead of creating new threads every time a thread is necessary [3]. A program can also put an upper limit on the number of threads in the pool, explicitly limiting the amount of parallelism and avoiding too many context switches.

In this section we present our research contributions in which we provided (reversed) architecture level modelling and analysis support for performance aspects related to threads.

In software architecture 'Views' are used to describe a set of specific concerns held by one or more system stakeholder. A Viewpoint describes the way a specific View is constructed. We developed an architecture viewpoint called Parallelism Viewpoint that describes the system's parallelism behaviour, which can be used to identify performance bottlenecks caused by excessive use of threads.

In the following subsections we describe our Parallelism Viewpoint and outline our approach to generate data graphs of the viewpoint. Subsequently, we describe how we used the graphs as part of the viewpoint to identify underused threads that can be replaced with a small sized thread pool.

Parallelism Viewpoint

The fundamental building blocks of the proposed Parallelism Viewpoint are: (1) a set of parallelism-specific concerns, (2) the corresponding stakeholders, and (3) a set of parallelism specific data graphs.

Among the parallelism-specific concerns we distinguish between (1) Time Allocation, i.e. the total time a thread is active during its life cycle, (2) Operation Distribution, i.e. the total number of operations performed by each thread during its life cycle, (3) Operation Types, i.e. the nature of the operations performed by threads, e.g. read and write operations, (4) Active and Non-Active Times, where a task is active if it is actually executing instructions and non-active otherwise, and (5) Execution Elements Management, i.e. the way a thread's life cycle is managed, e.g. a thread creation and deletion mechanism. Some of the corresponding stakeholders are software architects, developers and testers. The specific data graphs are Time distribution, Operation distribution, Operation types, Thread behaviour and Thread management. They are used to describe the aforementioned concerns. To illustrate the Parallelism Viewpoint, we discuss the Time distribution data graph which we developed for the electron microscope software system.

Time distribution data graph: An important task of an operating system is to allocate computer resources among competing execution entities. In many applications CPU time is a critical computer system resource and needs to be allocated appropriately. Threads which are the basic units of execution use their quota of CPU time to perform their tasks [4]. Threads enter the running state to use the allocated CPU time. A time distribution data graph illustrates the total active time for every thread in a system over a period of time. The active time is the amount of time a thread remains in the running state. Usually, during its life cycle, a thread enters the running state many times: we define Active Time as the time the thread remains in the running state for one instance. The total active time for a thread is the sum of all its active times. The time distribution data graph can be used to identify threads consuming no, or very little CPU time.

Viewpoint Data Graph Generation

In this subsection we describe the proposed approach for generating the afore-mentioned Parallelism Viewpoint data graphs. We first outline a parallelism specific execution meta-model and describe how to harvest information for its elements. Subsequently, we discuss how Parallelism Viewpoint data graphs are developed from this information.

Parallelism specific execution meta-model: A preliminary step towards generating the Parallelism Viewpoint data graphs is to identify the required runtime information. For this purpose, we explicitly categorise parallelism related runtime elements. Figure 4.1 shows the execution meta-model with these elements, which is a modified form of the meta-model proposed by Arias et al. [5]. The highlighted parts represent the modifications we made.

As shown in the top left part of the figure, the system execution starts with a user interaction. A user may perform multiple interactions with the system, which from the user's point of view logically fit together, e.g. focusing on a specific area of a sample or starting up the microscope. We refer to such a set of interactions as a scenario. The system performs the chosen scenario through a number of exe-cution steps, which represent a sequence of actions required to fulfil the intended interaction. Further, these steps are assigned to software components such as

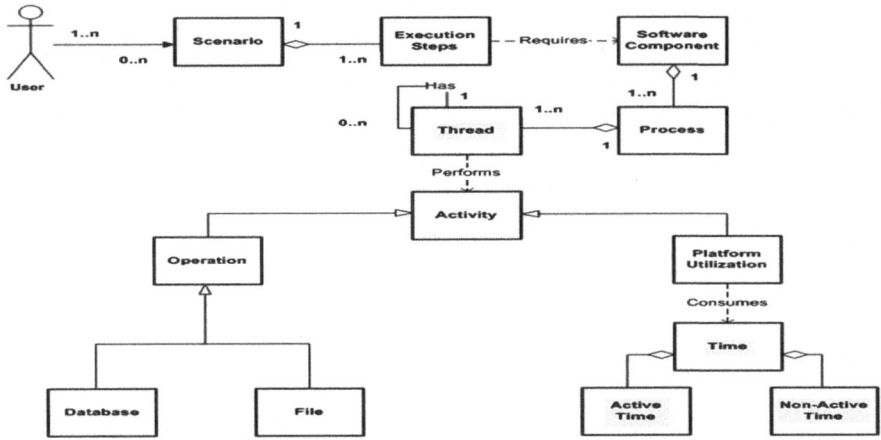

Fig. 4.1 Software system runtime information meta-model

source code modules. Subsequently, these software components are mapped to execution elements i.e. processes and threads. A process contains one or more threads. A thread may spawn multiple threads to hand out certain tasks to them.

The main thread activities are platform utilisation and I/O operations, where a thread is waiting for an operation, that is delegated to another system entity, to complete. These include interaction with e.g. cameras, the stage, the lens, etc. We considered platform utilisation and I/O operations as the main activities. However there are cases where a thread may perform other kinds of activities, such as some mathematical computations. Such activities are currently not specified in the meta-model. Finally, we explicitly include active and non-active times of a thread in the meta-model.

Data graph generation approach: In this subsection we describe an approach to harvest information about the meta-model's elements. Figure 4.2 shows the proposed approach. In this, we use process logs and logs maintained by the software system under investigation. A wide range of commercial tools are available to populate various kinds of information about processes. We logged our information using the Microsoft Process Monitor tool [6]. Information from logs is used to build a repository. Subsequently, the Parallelism Viewpoint data graphs are generated by using this information.

Thread Pool Analysis

Our Parallelism Viewpoint data graphs can be analysed to identify threads in the system that are suitable for thread pooling. For this purpose we proposed an approach consisting of two phases: an analysis of the thread pool, and validation.

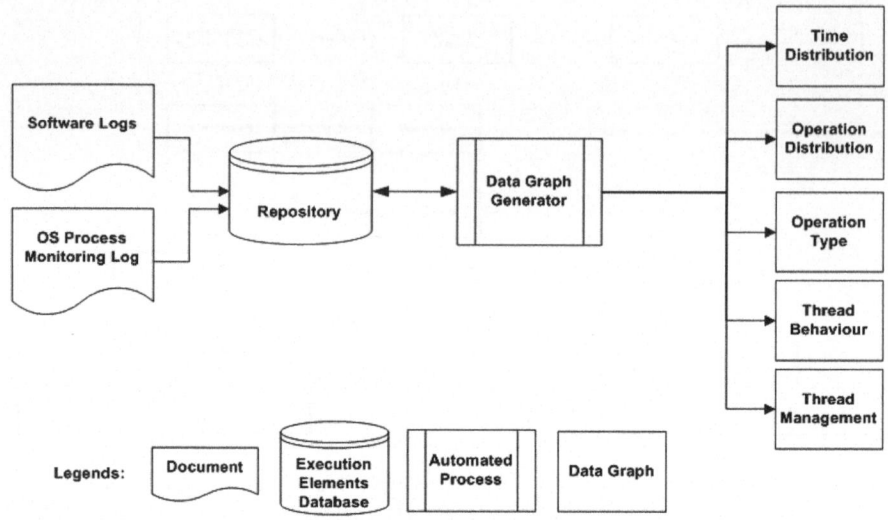

Fig. 4.2 Parallelism viewpoint data graph generation approach

In the first phase, based on a limit on the total time consumed and the total operations performed by a thread, a list of candidate threads is prepared. In the second phase, every thread in the list is assigned a priority (from 1 to 3) for its suitability for elimination. Priorities are assigned based on changes in thread behaviour in multiple scenarios. Our Parallelism Viewpoint thread behaviour data graph is used to identify changes in the behaviour. Priority 3 threads are most suitable to be replaced with a thread pool, priority 2 are less suitable, while priority 1 threads are not suitable as they have significant changes in their behaviour.

In a test we consecutively performed 3 different scenarios. We analysed the total number of tasks in the test and the tasks that could be candidates for thread pooling, given the criteria that the total time consumed and the amount of operations performed are limited. The test was carried out 10 times, giving different results depending on external circumstances, like the actual state of the controlled parts of the microscope, e.g. vacuum, lens, stage, etc.

Figures 4.3 and 4.4 show the candidate list and the priority list consecutively for the 10 analyses we performed. In Fig. 4.3, representing the first phase of our analysis, the horizontal axis represents the sequence number of the analysis whereas vertical bars represent the total and candidate threads in each of the ten analyses. The horizontal axis in Fig. 4.4, representing the result of the second phase, shows the sequence number of the analysis whereas the vertical bars represent the total number of threads for each priority. Only threads that were active in all three scenarios were considered. We can observe from these figures that in each analysis more than 50% of the total threads are candidate threads, the threads

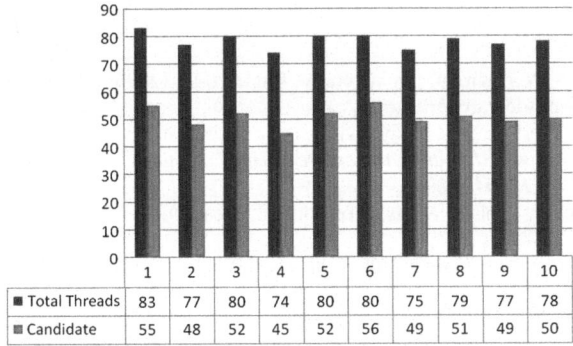

Fig. 4.3 Candidate list

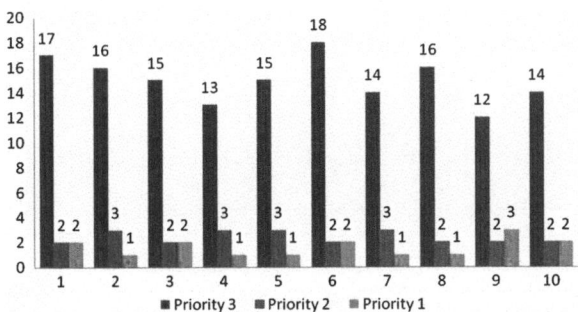

Fig. 4.4 Priority list

that are underused. And, the majority of these candidate threads are most suitable (priority 3) for possible replacement with a thread pool.

Conclusion

It is hard to tackle the performance bottlenecks of large software intensive systems. Software architecture and the use of models and architectural viewpoints can help to get a better view and grasp on the system, and to analyse what is really happening inside the system.

In hindsight, the problem we addressed, i.e. the large of amount of threads, did not turn out to be problematic at all. So, the gain of implementing thread pooling is negligible and consequently there is no reason to implement thread pooling. However, defining models for relevant software architecture aspects is in itself valuable. The way this was approached through the viewpoint method is difficult to assess based on this sole example.

References

1. TOGAF, The open group architecture framework (2006), http://pubs.opengroup.org/architecture/togaf8-doc/arch/toc.html
2. D. Xu, B. Bode, Performance study and dynamic optimization design for thread pool systems, in *Proceedings of the International Conference on Computing, Communications and Control Technologies CCCT2004* (2004)
3. Y. Ling, T. Mullen, X. Lin, Analysis of optimal thread pool size. ACM SIGOPS Oper. Syst. Rev. **34**, 42–55 (2000)
4. W. Stallings, *Operating Systems: Internals and Design Principles*, 6th edn. (Prentice Hall, NJ, 2008)
5. T. Arias, P. Avgeriou, P. America Analyzing the actual execution of a large software-intensive system for determining dependencies, in *Proceedings of 15th Working Conference Reverse Engineering WCRE* 2008, pp. 49–58
6. Microsoft Process Monitor (2011), http://technet.microsoft.com/en-us/sysinternals/bb896645. Accessed 8 June 2011

Part III
Automation and Control Functions

Chapter 5
Applications in Automated Microscopy

Abstract Automation of measurements and automatic analysis of samples on a transmission electron microscope enables a large variety of new microscopy applications suited for industrial needs. The consequences of these new possibilities are profound. The current methods for performing measurements and analysis should be considered carefully in order to find the optimal procedures and conditions needed to deliver the required accuracy or throughput. This chapter shows how to design an automated microscopy application using a model-based approach for one specific application, viz. the size analysis of a mixture of particles.

Keywords Microscopy application · Particle sizing · Particle detection · Experiment design

5.1 Introduction

Richard Doornbos and Sjir van Loo

Embedded Systems Institute, The Netherlands

As a result of the worldwide drive towards nanotechnology, there is an emerging market for tools and systems that allow us to observe and manipulate the ultra-small. Therefore there is a growing interest in nanoscale observation using transmission electron microscopes (TEMs) for industrial processes and applications. Typical industrial applications are chemical bulk production processes. In the last 10 years the production of materials like cement, latex and rubber, asbestos, pigments and whiteners, in the form of emulsions and molecular films

R. Doornbos and S. van Loo (eds.), *From scientific instrument to industrial machine*, 53
SpringerBriefs in Electrical and Computer Engineering,
DOI: 10.1007/978-94-007-4147-8_5, © The Author(s) 2012

has been undergoing a shift from uncontrolled batch processes towards strictly controlled continuous processes.

In biological research, a lot of attention is paid at the ultra-small scale to functions and structure of viruses, prions, proteins, etc. In this field, the safety of upcoming nanotechnologies should be investigated, better understood and handled, and possibly constrained by and captured into health regulations. Currently, nanoparticle toxicity is a hot research topic.

A second typical application goal lies in improving our understanding and building up knowledge about production processes; to optimise cost, product quality and production speed. Many chemical reactions are accelerated by catalysis and therefore catalytic reactions are a rich field of research.

Common key characteristics in these industrial applications are high demand for accurate and standardised analyses, and short analysis times. This immediately indicates the two required operation modes of a well-suited automated microscope: routine analysis for batches of product samples, and in-line real-time analysis in the production process. The first mode is typically suited to troubleshooting the production process, and performing research and development to improve the process. The second mode is typically used for direct process control.

A TEM has to perform three main types of observation functions in these new application areas. An elementary function is determining the size of nanoparticles. While apparently simple, in practice it proves very difficult to achieve both accuracy and precision in a short time using state-of-the-art systems. An extension to this function is shape determination, i.e. the creation of a three-dimensional representation of nanoscale objects, or aggregated structures. The third function is determining the chemical composition of particles in terms of atoms or molecules. Localising the exact elements with respect to each other is of crucial relevance for investigating and understanding the chemical reactions underlying the production processes.

The new applications impact on the system's requirements and are important drivers in the design of a dedicated electron microscope system. This means we also have to consider the overall workflow in which the nanoscale observation system is embedded, and take the associated requirements into account. It is readily seen that at least the following new functionality is required: automated specimen preparation, automated loading into and unloading out of the measurement system, automated measurement (imaging) procedures, and automated measurement data processing and interpretation. Also, the overall workflow must be cost-effective, have a high throughput and a short time-to-result. This means that the electron microscope must have high availability, high reliability, and provide reproducible results. Furthermore it should have the flexibility to cover industrial application needs, and be able to adapt to the expected changes in an industrial workflow.

The proposed functions and system qualities enable entirely new types of investigations that specifically exploit the strong features. To mention only a few:

- Making use of very large numbers of measurements, now enabled by automation, we can measure very small effects that hardly stand out from the noise or background signals. As an example, we mention the single particle analysis method, in which a three-dimensional reconstruction of protein complexes is derived from hundreds of images with randomly oriented proteins.

- Using the improved measurement stability provided by system control (e.g. drift compensation), we can measure dynamic effects much better. For example, the environmental TEM (ETEM) measures influences of a change in chemical environment on a given material.
- Using the improved reproducibility of the system, smaller differences can be measured. For instance the influence of some mechanical, electrical or chemical treatment can be assessed more accurately.

It is hard to find and tackle issues related to these new applications. To be able to identify and explore any issues, it is very beneficial to choose a set of microscopy applications as leading cases. The following section deals with one particular application investigated in the context of the leading cases (Concept Cars, see Sect. 3.1) in the Condor project: particle size measurements. This is a transmission electron microscopy application determining the sizes of a large number of particles in the range of a few to hundreds of nanometres. As mentioned, this seems a relatively simple case and similar to what is being done in optical and *scanning* electron microscopy (using SEM), but it contains many challenges in the design of the measurement procedure (experiment design). The model-based investigation considers finding and sizing nanoparticles, optimising measurement for speed, and finally estimates throughput improvement when going from human to automated nanoparticle sizing.

Acknowledgments The authors like to acknowledge Yuri Rikers from FEI for sharing his experiences in electron microscopy.

5.2 Throughput Maximisation of Particle Size Measurements

Wouter Van den Broek

EMAT, University of Antwerp, Belgium

Context

Measuring the size of a large number of particles by an operator is time consuming and expensive. Furthermore, there is a risk of involuntary 'cherry picking', for example when a few big particles are neglected in favour of smaller ones, thereby distorting the radius distribution. Automation improves the throughput, allows continuous measurements, 24 h a day, and can be more reliable.

In this section we show how automation can yield a higher throughput. There are three challenges:

- Automatic particle detection: the difficulty in recognising the particle of interest in an unstructured background.
- Automatic radius measurements: determining the particle's size with sufficient accuracy.
- Throughput optimisation: finding the optimal measurement conditions for a particular sample.

We present an algorithm that finds a large percentage of the particles in an image without human involvement. When the particle positions are found, the radius will be measured by fitting a model to the image intensities. Throughput is increased dramatically if the appropriate spot size is selected. We show a method that automatically finds the optimal spot size using measured data.

At the dimensions under investigation, a few to hundreds of nanometres, no objective size standards exist. This is exemplified in a reference paper on size measurements [1], where, depending on the measurement techniques used, the United States' National Institute of Standards and Technology arrives at widely different estimates for the diameters of gold nanoparticles. By incorporating a model of the chosen measurement technique in the estimation process, the inherent systematic errors can be taken into account, resulting in a more reliable estimate.

Image Model

To address the above challenges, we need detailed models describing the creation of the electron microscopic image. In their most general form, too many parameters come into play, making these problems prohibitively difficult. We therefore need to restrict the problem and use some reasonable simplifications.

In this section it is assumed that the particles are imaged by high angle annular dark field scanning transmission electron microscopy (HAADF STEM). In this mode, an electron probe is raster scanned over the specimen. For every probe position only the electrons scattered to high angles are recorded. The resulting image is a convolution of the projected specimen function with the probe profile, see Fig. 5.1.

The particles lay without overlapping on an amorphous support of low atomic number (typically carbon). The particles have shapes that are reasonably well approximated by spheres. For the probe profile a Gaussian function is assumed. The image of a particle can then be described as the projection of a sphere, convolved with the (Gaussian) probe profile, superimposed on a background of relatively low intensity.

The radius of the particles is generally assumed to follow a log-normal distribution [2], see Fig. 5.2. A log-normal distribution is characterised by its geometric average

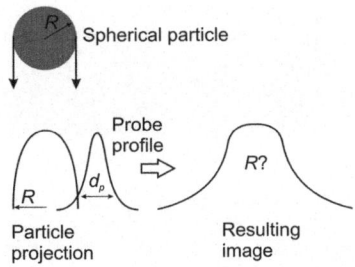

Fig. 5.1 Image formation process in HAADF STEM. The projection of the spherical particle with radius R is convolved with a probe profile with width d_p

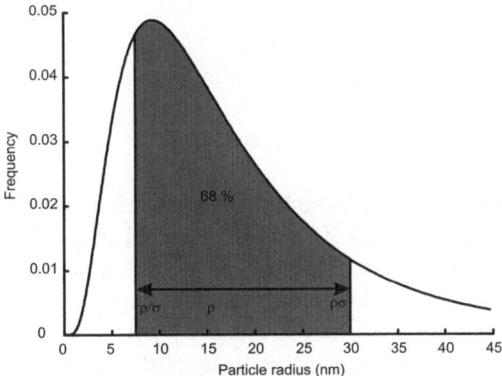

Fig. 5.2 Log-normal particle radius distribution, with a geometric mean ρ of 15 nm and a geometric standard deviation σ of 2. About 68% of the particles have a radius between ρ/σ and $\rho\sigma$ (7.5 and 30 nm, respectively), and 95% have a radius between ρ/σ^2 and $\rho\sigma^2$ (3.75 and 60 nm, respectively)

ρ and its geometric standard deviation σ. In the remainder of this section, ρ and σ are referred to as 'the average' and 'the width' of the particle radius distribution.

Although this image model rules out elongated particles and high resolution images, it is general enough to be used as a test case that covers a broad range of practical applications.

Automatic Particle Detection

Judging from Fig. 5.3, the contrast in the HAADF STEM image is high and one could be tempted to identify the particles solely by their grey values, known as the thresholding method. However, choosing the appropriate threshold without any human involvement is a non-trivial task for two reasons. Firstly, the intensity of the particles relative to the support is not known a priori and can vary across the

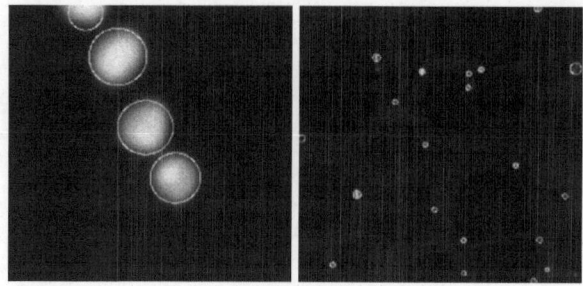

Fig. 5.3 Results of the particle detection. *Left*: Touching particles are correctly identified. *Right*: Most particles are found on a complicated background. In the upper right corner, two particles are identified as one

sample. Secondly, the radii of the particles follow a log-normal distribution and therefore can have a large range, resulting in widely varying image intensities.

A procedure was developed that does not involve thresholding, but makes use of the local variance of the image. To detect if a certain region contains a particle, one can draw a square on it that is centred on the centre of mass of the grey values. The square surrounds a particle if the local image variability stops increasing when the square's sides are increased. This has been implemented in a Matlab script.

The detection algorithm finds most of the particles in practice. The fact that real particles are not exactly spherical seems of no importance. When two particles of only a few pixels are touching they are usually detected as one. If the touching particles are large compared to the pixel size, both are often identified correctly. See Fig. 5.3.

Adequate sample preparation is needed. In the case of standard holey carbon film support, the regions between the holes are often wrongly identified as particles. It is therefore advisable to use a carbon film without holes. It is furthermore important that the particles are uniformly spread over the support with only a little overlap, whilst still having enough particles in the field of view at the highest magnifications.

Automatic Radius Measurement

Once the particle position has been found, the image model is fitted to the local region to estimate the particle radius. This model has 5 parameters: the radius of the particle, the x- and y-positions, and the intensities of particle and background. The width of the probe profile is assumed as prior knowledge. By writing the linear parameters as a function of the nonlinear parameters [3], the number of variables reduces to 3.

Tests on both real and simulated images showed that the fitting procedure is robust and always converges. The probe width is often not known exactly, but it is small in comparison with the particle radii. Therefore, assuming that the probe width is zero results in a negligible error.

The number of particles measured with an automated procedure can be far too high to be checked manually for reliability. Therefore, an automated checking procedure is necessary. If Poisson noise is assumed, a figure of merit (FOM) can be derived that describes how well the model fits the measurements. If the FOM is suspiciously low, something went wrong–for example, the particle detection might have identified two partly overlapping particles as one–and the result must be rejected. However, in practice the FOM does not perform well because the model is never exact. The particles are never exactly spherical and the background not exactly uniform, causing too low a FOM even in the case of a good fit. Although this particular choice of FOM did not work out in practice, the concept of a FOM remains critical in an automated environment. The reason is that the number of investigated particles is too large to allow for manual quality control of the estimation procedure.

While testing the estimation procedure on experimental images and during the concept car demonstration, we were surprised to notice that the estimation works well even if the particle was sampled by only 3×3 pixels, which is much less than a human operator would typically choose. This increases the throughput considerably, see the last subsection.

Spot Size Selection

It is assumed that the customer wants to know the particle radii with a certain pre-specified precision σ_R. In this subsection we investigate which spot size yields the highest throughput T, while meeting the required precision. T is defined as the number of particles recorded per unit of time, and therefore is inversely proportional to the recording time per unit of area. It is necessary to formulate an explicit relation between precision and throughput, so the user can make the trade-off between both. This relation will contain microscope parameters and parameters characterising the sample.

The relation between precision and throughput can be found by using Bayesian statistical experimental design [4]. In Ref. [4] we show that $T = \alpha \, \sigma_R^2$. The factor α is a complex function of the microscope parameters and the parameters characterising the sample, but the only tuneable parameter is the spot size. The spot size maximising α will be denoted the Bayesian optimal spot size d_B.

The Bayesian optimal spot size d_B is a function of the sample parameters, i.e. the average ρ and width σ of the particle radius distribution, and the atomic number of the particles. For narrow radius distributions (small σ), d_B approximately equals ρ and is independent of the microscope parameters. For wide radius distributions (large σ), d_B is independent of the particle radius distribution and equal to 1.25 times the geometrically limited spot size d_g, which is solely determined by the microscope parameters. The optimal spot size for radius distributions of intermediate width is situated between these two values and has to be

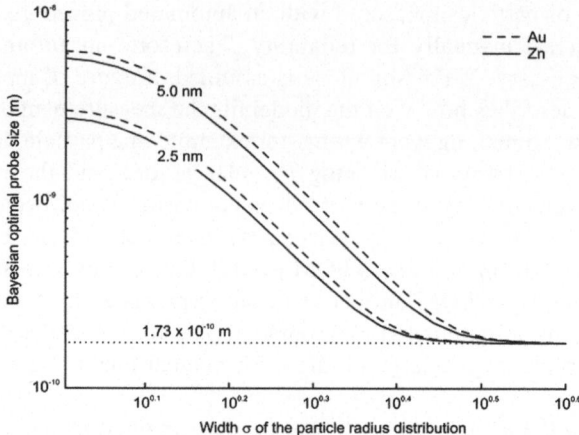

Fig. 5.4 Optimal spot size as a function of the width σ of the particle radius distribution for four cases: gold (^{79}Au) particles with an average radius ρ of 2.5 and 5.0 nm, and zinc (^{30}Zn) particles with an average radius ρ of 2.5 nm and 5.0 nm. See [4] for the microscope parameters

determined with a numerical optimisation. Figure 5.4 also shows that dependence of d_B on the atomic number is negligible.

These results suggest that a speed-up of the size measurement procedure is possible by means of an adaptive approach. A broad radius distribution can be interpreted as having no prior knowledge at all about the sample. This means for an unknown sample it is best to start with a probe width of 1.25 d_g. As the measurement progresses and knowledge about the radius distribution increases, the spot size can be adapted continuously.

Comparison to a Human Operator

In this subsection we discuss the throughput of an automated system compared to that of a human operator. This is necessarily an approximate endeavour, because operators' actions and customs are not standardised and can vary widely. Therefore, assumptions are made that the author feels are reasonable. Because the result is sensitive to these choices, care should be taken to justify them.

For this comparison to make sense, the precision of both size distribution estimates has to be equal. Therefore, the microscope settings must be chosen so that both estimation procedures can make use of the same amount of electrons to estimate the radius of a particle of average diameter.

Three factors work together to cause the difference in throughput. Firstly, the automated procedure needs fewer pixels for the smallest particles than the human operator does (3×3 vs. 10×10 pixels). Secondly, because the automated procedure takes the probe width into account, it can use a wider probe with more current.

Thirdly, to ensure the human operator and the automated procedure produce the same precision, the operator uses the same dose for a *profile* over a particle of average size as the automated procedure uses for the *entire image* of that particle.

Preliminary calculations show that throughput increase by the automated procedure over the manual procedure depends heavily on the shape of the particle radius distribution. It varies widely between factors of 10 and 1,000, which–given our assumptions—signifies that the automated system will always outperform a human operator.

Conclusion

The automated particle detection and radius measurement have been validated experimentally in the Concept Car prototype (Sect. 3.1). The spot size selection has primarily been validated through computer simulations, although it was also implemented in the Concept Car.

Combining the automated particle detection and radius measurement with the selection of the optimal spot size increases the throughput significantly, as discussed in the last subsection. Furthermore, the microscope can be operational 24 h a day, no tedious manual particle measurements are needed, and the microscope can be operated in conditions that are not attractive to human operators.

In this report we have used a different optimisation approach as the throughput at constant precision has been maximised, not so much by increasing the information per electron, but rather by increasing the information per second.

Acknowledgments The author likes to acknowledge Max Otten from FEI for sharing his experience in probe formation.

References

1. V.A. Hackley, J.F. Kelly, Report of Investigation, Reference Material 8011, Gold Nanoparticles, Nominal 10 nm Diameter. National Institute of Standards and Technology.
2. W.K. Brown, K.H. Wohletz, Derivation of the Weibull distribution based on physical principles and its connection to the Rosin-Rammler and lognormal distributions. J. Appl. Phys. **78**, 2758–2763 (1995). doi:10.1063/1.360073
3. W.H. Lawton, E.A. Sylvestre, Elimination of linear parameters in nonlinear regression. Technometrics **13**, 461–467 (1971)
4. W. Van den Broek, S. Van Aert, D. Van Dyck, Throughput maximization of particle radius measurements through balancing size versus current of the electron probe. Ultramicroscopy **111**, 940–947 (2011). doi:10.1016/j.ultramic.2010.11.025

Chapter 6
Focusing Control

Abstract Creating sharp images on a transmission electron microscope involves extremely precise setting of the electromagnetic focusing elements in the system. In the context of focus control it is useful to distinguish coarse and fine grain setting regimes, as these have different requirements. Coarse grain focus setting by electromagnetic lenses is difficult, mainly due to hysteresis caused by the iron constituents of the lens. This chapter describes an approach to control and optimize the coarse-grain setting of focus, including an explicit trade-off analysis of speed versus accuracy. In fine-grain setting the challenges reside mainly in image-based multi-parameter optimization of image sharpness, which is influenced by both lens and stigmator settings. This chapter also describes methods to optimize for focus and astigmatism, and their application in measurement procedures.

Keywords Defocus control · Image sharpness · Electromagnetic lens · Hysteresis · Trade-off analysis · Astigmatism · Sharpness function · Multi-parameter optimization · Nelder-Mead simplex method

6.1 Introduction

Richard Doornbos and Sjir van Loo

Embedded Systems Institute, The Netherlands

The automated electron microscope's key function is creating images with sufficient quality for the subsequent automated image processing chain, as discussed in the general Introduction. The meaning of 'sufficient quality' is determined by many factors, e.g. the type of sample, the measurement conditions, the application goal

R. Doornbos and S. van Loo (eds.), *From scientific instrument to industrial machine*,   63
SpringerBriefs in Electrical and Computer Engineering,
DOI: 10.1007/978-94-007-4147-8_6, © The Author(s) 2012

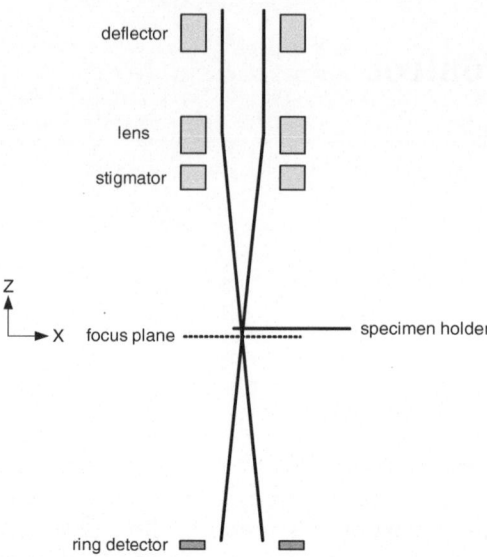

Fig. 6.1 Simple model of a STEM

and robustness of the image processing algorithms. For our purposes, the signal-to-noise ratio and 'image sharpness' are good metrics for assessing the image quality. To understand how things are related, we first explain a simple model of how a scanning transmission electron microscope (STEM) creates images.

In Fig. 6.1 we see the structure of a STEM, reduced to its essentials. The electrons enter from the top (the beam bounds are indicated by the black lines). Their paths are influenced by several components in order to pass through a small measurement area. The lens causes the beam to converge, with a beam width as small as possible in the focal plane. When electrons impinge on a specimen, they interact via a multitude of physical phenomena, causing energy loss and/or direction change. The transmitted electrons detected by the ring detector (high-angular annular dark field, HAADF) have all experienced a direction change. The detected signal is proportional to the amount of material and the beam intensity in the interaction volume. To create an image, the very fine beam spot is scanned (bending the beam using the deflectors) in two dimensions across the specimen area, and correlated with the signals from the detectors (much like a television tube). When a thin specimen is located exactly at the focal plane, the maximum beam intensity is achieved, leading to optimum signal-to-noise ratio, and the minimum beam width leads to the smallest interaction volume. In other words, the specimen is 'in focus'. Obviously, when the specimen is not in the focal plane, the image will exhibit lower signal to noise ratio, and will be perceived as 'unsharp'.

A complicating factor is the *astigmatism*. The focus plane for the two lateral directions (X-direction and Y-direction in Fig. 6.1) are slightly different in practice, leading to typical, but unwanted, distortions in the image. To compensate for this difference, *stigmators* are present that change the direction of the electron beam in one direction only.

In a TEM the deflectors and lenses are usually electromagnetic, thus their electric coils have to be driven with a certain current. Therefore, setting the focus plane in scanning mode is performed by setting the currents in the lens assembly.

In contrast with optical lenses that have a fixed focal length, these electro-magnetic lenses are continuously variable. When we consider changing the focusing strength of lenses, we can identify three essentially different regimes: (1) setting to a certain focus value, starting very far from the optimum (no usable image); (2) re-focusing, when already being close to the optimum (at least something is visible in the image); and (3) keeping the focus at the optimum within certain bounds (sufficiently good image). These three regimes require different approaches due to a precision-critical instrument's complicating factor that it must also use its generated image for 'sensing' and control purposes (see Sect. 2.2). The image must be used for determining the effect of a change in e.g. lens strength by measuring the sharpness. Based upon the detected change in sharpness, the next step in a correction or optimisation procedure can be determined.

But why can't we just set the current to the desired value, and position the focus plane exactly where we want it? The microscope operates at the cutting-edge of what is technically possible. Therefore, it is inherently highly sensitive to (variations in) internal and external phenomena. Disturbances from outside the microscope, like floor vibrations, magnetic fields, air pressure changes, and air and cooling water temperature variations, cause a variety of effects on the microscope's behaviour. The effects can, in principle, be modelled using first principles, but the structural complexity of the instrument makes this a gigantic undertaking. The internal disturbances can be separated into active and passive disturbances. What we consider as active disturbances are in fact side-effects: when the system performs a stage move, as a side-effect the images often go out of focus. This is due to either imperfect horizontal alignment of the stage with respect to the electron beam, or a height difference in the specimen. When the magnification is changed, the image often goes out of optimum focus and becomes stigmatic. Many more system induced changes exist that exhibit these types of side-effects. On the other hand, passive disturbances are usually physical relaxation processes like the dissipation of internally generated heat that causes temperature changes in critical components, e.g. the sample holder. Another example is the relaxation of stage elements. Behaving like loaded springs they release their built-up tension at unpredictable moments, leading to position drift (most noticeable right after a move). The magnetisation of the lens and stigmator iron has the same type of relaxation process, although at somewhat shorter time scale.

A separate phenomenon is *hysteresis*. This is essentially 'system path dependence': the behaviour of a system depends on the (hidden) states of its constituents, which are influenced by their history. A typical example of hysteretic behaviour in the electron microscope is visible in setting the magnetic lens strength. The source of the hysteresis effect is a change in iron magnetisation, created by altering the directions of the discrete magnetic domains at micro-scale.

Predicting the behaviour of the lens very accurately has proven extremely difficult. A mechanical example of hysteresis can be found in the backlash of the stage (see Sect. 7.2).

Now that we more know about the issues, let us resume our discussion about the classes of setting microscope system properties. As stated, we identified three types of system property setting which need a different approach.

1. Coarse-grain setting

Used to make major changes, i.e. significant fractions of the full range. In the magnetic lens coil we can see current changes from 0% to nearly 100%. This type of setting appears frequently in normal microscope use, e.g. at start-up, in a magnification change in TEM mode, and in applications like the slice-and-view application in a dual-beam SEM (in which the ion beam cuts material and the electron beam is used to view the result). The purpose of this setting is to 'get close' to the optimum (say within 1%), knowing that setting to the exact optimum cannot be achieved in one step. An important requirement is to achieve the new setting fast. This requirement leads automatically to making a trade-off between time and accuracy.

2. Fine-grain setting

Used for smaller changes. For our lens case we consider current changes around and below 1%. These smaller changes to the system occur very frequently, for instance to make corrections, or compensate for unwanted side-effects or drift of system properties. In an automated microscope these appear e.g. after a magnification change (in STEM mode) or after a stage move, when the optimum focus has to be restored. The purpose of this type setting is to get 'from close, to optimum'. Again one has to agree on an acceptable residual value, which depends on the acceptable error in the final measurement result. In the focusing example we assume the lens current residual to be in the order of 0.001%. The optimisation speed is also an important parameter here as the procedure is often repeated, therefore it will contribute significantly to the overall duration.

3. Fine-grain control

Slightly different from the previous, this type of setting has a continuous or repetitive nature. It is used in intermittent compensation or in control methods. In an automated electron microscope we see this type appearing for instance during payload acquisition when the system actively keeps the image sharp (in focus). The purpose of this type is a typical control objective: 'remain very close to optimum', having residuals in the same range as in the previous case. Obviously, in this case there is also a trade-off between time and accuracy, but the robustness of the control system against disturbances is considered more important.

In the following sections we elaborate on the above types of system property settings and their issues. In Sect. 6.2, type 1 is elaborated on extensively for an electromagnetic objective lens. In the subsequent section the focusing of the electron microscope using two methods that fall in the type 2 category is described: a three-parameter search optimisation and a fast one-parameter correction method. Both advanced functions were implemented and essential elements of the Concept Car (see Sect. 3.1).

6.2   Coarse-Grain Setting of the Magnetic Lens Focus: Dealing with Hysteresis

Patrick van Bree, Nelis van Lierop, Paul van den Bosch, Mart Bierhoff[*] and Seyno Sluyterman[*]

Control Systems, Eindhoven University of Technology, The Netherlands
[*]FEI Company, The Netherlands

Industrial Problem

For decades, electron microscopes were tools almost exclusively used by researchers studying material properties. The result was mainly a high-quality, highly magnified image for analysis and interpretation. However, next to the research market new markets have evolved in which the key system qualities are now expressed in *accuracy, precision* (e.g. standard deviation of the estimated size of particles) and *throughput* (number of images per time unit) for a certain microscopy application.

One of the system's throughput limiting factors is setting the focus. As the dynamics of focusing is largely unknown, the solution today is just to wait for the system to converge to constant values, which obviously limits the machine's throughput. Consequently our challenge lies in understanding the dynamic behaviour of the focusing subsystem of the electron microscope, which is currently neglected. Our aim is to apply control to decrease the settling time to improve the throughput.

Analysis

Focusing an electron beam is performed by a magnetic field created by an electromagnetic lens. This lens consists of a coil (or set of coils) surrounded by a solid ferromagnetic yoke (NiFe). The geometry of the yoke, in combination with the current running through the coil, determines the magnetic flux density distribution. Figure 6.2 shows an example of the geometry and magnetic flux distribution at the optical axis in steady state for different input current amplitudes. The focal distance of the lens is a function of the magnetic flux density at the optical axis and can be varied by changing the input current. Therefore, the relevant quantity to

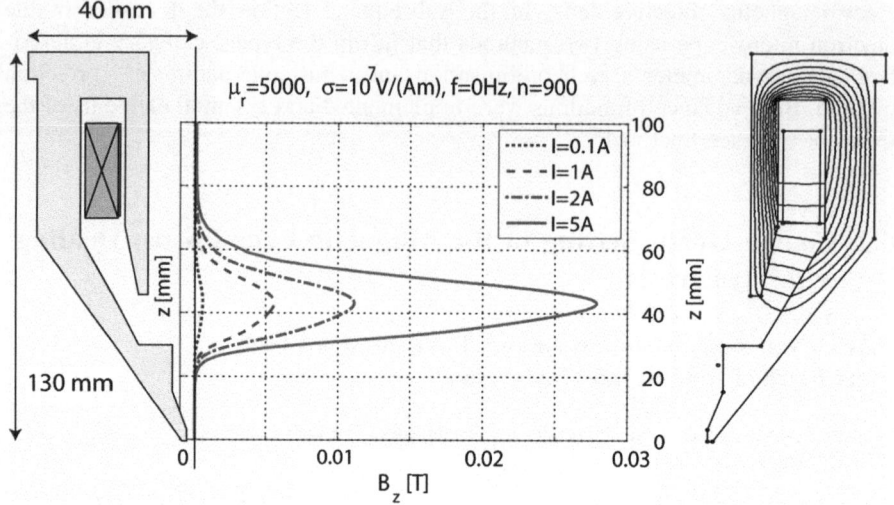

Fig. 6.2 Lens geometry (**a**) and magnetic flux density at the symmetry axis for different current amplitudes (**b**), and magnetic field line distribution (**c**)

control the optical focus is the magnetic flux density at the optical axis, via the input lens current. The relation between input current and magnetic flux density distribution is not well understood as it involves hysteresis, eddy currents and (local) saturation.

Control

One of the important characteristics of a control system is its ability to deliver high-performance system behaviour even when knowledge of the process is limited. If an accurate measurement of the output is available, an appropriate controller can suppress both model inaccuracy and disturbances. Classical controllers such as PI or PID are often used with satisfactory results. So, small variations in the process are completely acceptable. We can distinguish two different situations:

- **Accurate measurements possible:** If an approximate process model is available, a stabilizing controller with high gain can be designed. Even in the presence of disturbances, this feedback controller guarantees the output will follow the reference signal despite variations in the process. Accurate or approximate knowledge of the process and accurate measurements are a necessity to design these controllers.
- **No measurements are possible:** If there are no disturbances *and* the process is accurately modelled *and* the model of the process is invertible, feed forward control is sufficient to realise good control. However, these requirements are quite demanding.

In both situations we need at least an approximate model to apply control successfully. Starting with the physical laws that describe the basic dynamic characteristics of the process, a first qualitative model can be derived. For our purpose we want to have a validated model that provides reliable quantitative results. Consequently, parameter estimation schemes and measurements have to be used to obtain proper quantitative values. Stated roughly: *measurements can yield knowledge about an unknown process.* But designing an experiment requires quite a lot of knowledge about the intricacies of the process (e.g. bandwidth, signal-to-noise ratio, linearity). Therefore, on the other hand: *experiment design requires some knowledge about the unknown process.*

Measurements on an Electron Microscope

The focusing process proved extremely sensitive to changes in magnetic field strength: only 0.01% of the current range yields acceptable images (e.g. both images in Fig. 6.3 are acceptable as input for an autofocus procedure, see Sect. 6.3). Note that this means 99.99% of the range does not yield usable images.

Focusing (image sharpness) is a single-valued function of the magnetic flux density distribution. However, the magnetic flux density is a multi-valued function of the lens current: hysteresis, i.e. the history of the input current, determines which specific point out of the available values is the resulting one, see Fig. 6.3. Therefore the reproducibility of settings is complicated significantly by the hysteresis effect.

To ensure we work with realistic criteria and assumptions these effects have to be measured on a state-of-the-art scanning electron microscope (SEM). However, this requires dedicated hardware and software as we want to find the dynamic, and not the static relation between magnetic lens input currents and image sharpness. An FEI microscope was extended with a rapid-prototyping system for controlling the objective lens, and a dedicated synchronisation system to guarantee synchronisation of the time between control signals and capturing images.

Our measurements clearly confirmed that hysteresis in the yoke introduced a considerable problem for accurately reproducing image sharpness. One example is shown in Fig. 6.3 (right), where completely different image qualities are obtained with the same lens current.

More importantly, we were unable to find models that fitted our measurements and which were accurate enough for our purposes. Therefore, we still do not have sufficiently accurate models for which the effects of hysteresis could be predicted and used in a feed forward way to control the lens current.

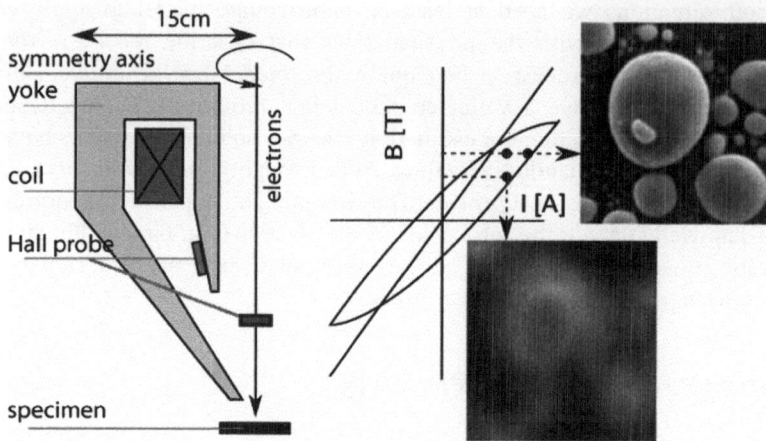

Fig. 6.3 *Left*: cross-section of a magnetic electron lens. As a first approximation the lens can be considered circle symmetric. Note the two locations for the Hall probe. *Right*: influence of hysteresis on image-quality. The same input current results in two completely different images (**a**) and (**b**)

Measurements on a Dedicated Set-Up

From the experiments with the FEI microscope it became clear we needed extremely accurate measurements of the magnetic flux in the yoke. Magnetic field sensors can provide the required accuracy, but they cannot be installed without major investments. Therefore we created an experimental set-up with only the essential components: the magnetic lens and its ferromagnetic yoke, and a data acquisition and rapid prototyping system. To meet the specifications for electron microscopy applications, we selected the most accurate, wide-range, high-bandwidth magnetic field sensors currently available, and an over-dimensioned power amplifier. The sensor (Hall probe) was placed within the lens geometry as shown in Fig. 6.3. With this set-up, control performance can be evaluated experimentally. Note that, instead of image quality (sharpness), the performance is now based on measured behaviour of the magnetic field.

The experimental set-up allowed us to investigate different strategies for fast and highly accurate set point control of the magnetic flux density within electromagnetic lenses. The following controller structures have been evaluated.

A: Feed forward without magnetic flux sensing. The controller scheme, shown in Fig. 6.4, is used when there is no sensor available.

B: Feedback control with an optimal sensor placement. The controller scheme for an optimal sensor position, i.e. at the location of the magnetic field lens, is shown in Fig. 6.5. This is obviously not a usable set-up as the sensor blocks the beam completely.

C: Feedback control with a constrained sensor position. When the sensor is at a location close to the optimal one, the controller scheme can be modelled as

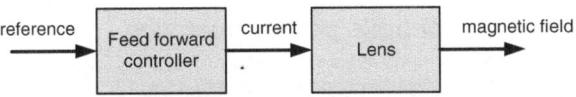

Fig. 6.4 Feed forward control scheme A

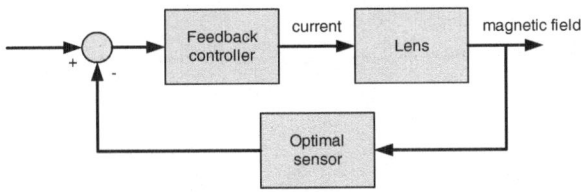

Fig. 6.5 Feedback control scheme B

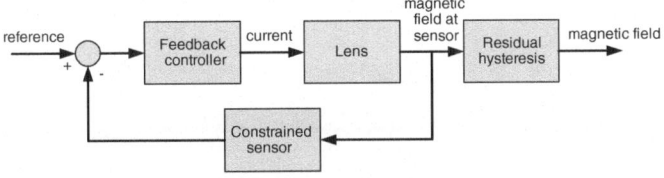

Fig. 6.6 Feedback control scheme for a non-optimal sensor location C

shown in Fig. 6.6. Note that a residual hysteresis relation in the magnetic fields exists between the location of the sensor and the optimal location.

Electron microscope operators use a procedure, named *initialisation*, when very accurate focus setting is required. Initialisation forces the state of the magnetic yoke to a reproducible value by using multiple periods of an excitation with currents covering the complete range. This erases the yoke's memory and therefore reduces the influence of initial conditions. We investigated the feed forward initialisation procedure applied to both feed forward and feedback schemes.

Results

For making trade-off analysis easier, we created a controller performance map that shows the transition error versus the transition time, see Fig. 6.7. The performance of feed forward, feedback, and feed forward initialisation strategies are depicted in one diagram for comparison.

Based on first-principle derivations, it was shown that the desired accuracy for fine-grain control is about 10^{-5} of the complete range. The accuracy, required to guarantee overall machine performance based on image-based defocus control (coarse-grain control), is 10^{-4} of this complete range. This is shown as an upper bound on transition error in Fig. 6.7. The lower bound on transition error is derived from physical constraints given by actuator and sensor limitations.

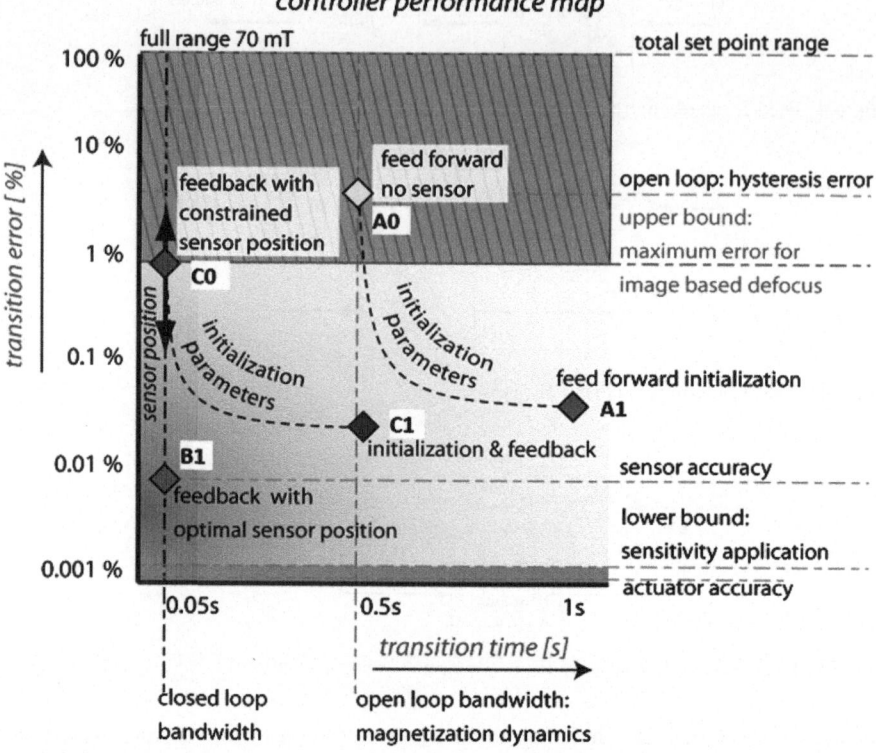

Fig. 6.7 Experimentally obtained performance map of transition error versus transition time for the three schemes (*A, B, C*) with (*1*) and without (*0*) initialisation procedure. The *dashed lines* indicate expected performance of the initialisation procedure as a function of the initialisation parameters. The *arrows* at *C0* indicate that moving the sensor influences performance in terms of transition error, but the low transition time is maintained

Using our dedicated set-up the baseline was determined experimentally: the transition error in feed forward is 5 times higher than the maximum allowed by application requirements. Also, the transition time in this setting was estimated to be 0.5 s, which obviously should be as small as possible for a high throughput. This baseline is indicated as A0 in Fig. 6.7.

Experimental validation of our newly developed controllers showed that several advantageous strategies are now possible.

- While keeping an equal transition time, the transition error can be decreased by over 150 times using magnetic flux density sensors and feed forward initialisation (from A0 to C1).
- The transition time can be decreased by a factor of 10 from 0.5 s to 50 ms based on feedback control. The improvement in error reduction is 5 times (from A0 to C0).

- In feed forward without any online magnetic flux density sensing, the transition error can still be reduced by 100 times using feed forward initialisation. The price paid is in initialisation duration. For the performance mentioned, initialisation increases the transition time of 0.5–1 s (from A0 to A1).

The performance limiting factors of feedback control are due to the constraints on the sensor position and the hysteresis effect (indicated by the arrows near C0). The sensor position within the lens geometry, combined with spatial dependence of hysteresis effects, can cause a residual problem: again hysteresis limits the performance, but on a smaller scale. Our solution is to control the relation between the magnetic flux density at the position of the sensor and the magnetic flux density at the relevant electron optic interaction volume in feed forward. Feed forward strategies work because the system is an open-loop stable system, the influence of any external disturbances is insignificantly low, and the lens current can be controlled with a high bandwidth, a wide dynamic range and an extremely high resolution (~ 1 μA).

Our research yields the requirements for an optimal initialisation excitation (along the curve A0–A1, and C0–C1), for an example see [1]. The feed forward initialisation approach is evaluated at the electromagnetic lens set-up. Our main conclusion is that, due to eddy currents and magnetic skin effects, the maximum frequency usable for initialisation of the FEI electromagnetic lens is about 50 Hz.

Conclusion

The presented control approach opens up many possibilities for improving speed and accuracy of the lens magnetic flux density setting procedure. It allows a fast transient to a required lens focus value (coarse-grain setting) such that the hand-over to image-based focusing (fine-grain setting) can work successfully. Furthermore, the consequences of switching from coarse-grain to fine-grain setting are much better understood. Designers can now make trade-offs between accuracy and transition times, in the context of various other trade-offs such as cost and design effort.

6.3 Fine-Grain Setting of the Lens Focus: Dealing with Defocus and Astigmatism

Maria Rudnaya

CASA, Eindhoven University of Technology, The Netherlands

Fig. 6.8 A photo and two synthetically generated images. From *left* to *right*: in-focus image without astigmatism, out-of-focus image without astigmatism, out-of-focus image with astigmatism

As explained in the introduction, focusing is defined as the act of making the image as sharp as possible by adjusting the lens. The adjustable parameter of the microscope which controls the current through the magnetic lens is known as defocus. Due to the presence of astigmatism the image cannot be totally sharp, and has different amounts of blurring in different directions. Figure 6.8 shows a photograph and the simulation of defocus and astigmatism effects. The stigmatic image on the right is not just unsharp. We can observe a stretching in a particular direction (in this case the horizontal).

In general we cannot estimate defocus and astigmatism using only one image from a sample with unknown geometry. Thus more images have to be recorded. Many existing autofocus and some astigmatism correction methods are based on a sharpness function, a real-valued estimate of an image's sharpness (an overview of existing sharpness functions can be found in [2]). For a through-focus series of images, the sharpness function is computed for different defocus values. An ideal sharpness function shape is shown in Fig. 6.9a. The image at optimum defocus is sharp and the sharpness function reaches its maximum. An image at a different defocus value is perceived out of focus.

Ideally sharpness functions have a single optimum (maximum or minimum) located at the in-focus image. In reality the sharpness function may have local optima due to lens astigmatism and noise in the image formation as well as the specimen's geometry, see Fig. 6.9b. Therefore our challenge is to find the global optimum, which provides the best image.

In the following subsections we describe the automatic optimisation of an electron microscope using two methods that fall into the fine-grain setting category. We first look at a simultaneous defocus and astigmatism correction method, based on the three-parameter optimisation of a sharpness function. In the second subsection we describe a fast one-parameter correction method. This is based on the interpolation of the derivative-based sharpness function with a quadratic polynomial.

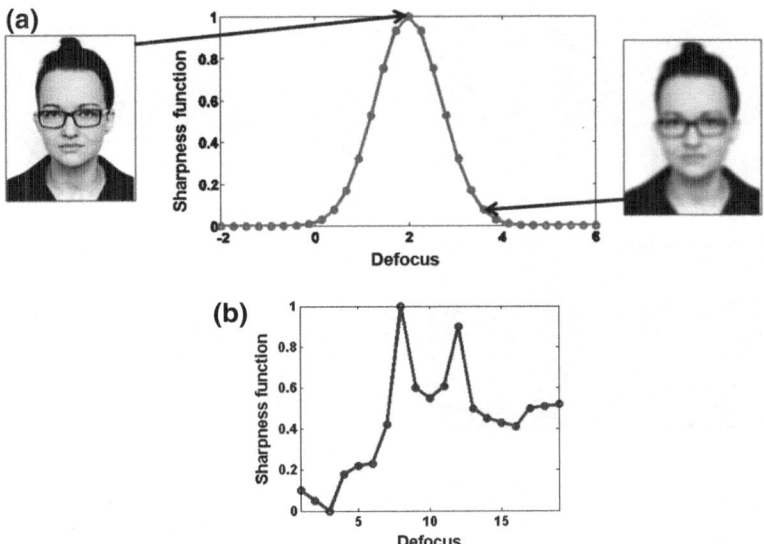

Fig. 6.9 a Ideal sharpness function. This function reaches its optimum at the in-focus image. **b** Typical sharpness function for a realistic specimen

Simultaneous Defocus and Astigmatism Correction

Determining the sharpness value is much faster than recording an image. Also, repeated recordings can damage or destroy the sample. Therefore defocus and astigmatism must be corrected with a minimum number of image recordings. The sharpness function simulations described in [3] show that it is important to adjust the microscope's defocus and astigmatism at the same time, due to mutual dependencies. The simultaneous defocus and astigmatism correction method we developed is based on the three-parameter optimisation of a sharpness function.

The Nelder-Mead simplex optimisation method is designed to find a local optimum of a function, in particular the variance-based sharpness function [3]. It makes no assumptions about the shape of the function. A detailed description of the Nelder-Mead simplex method can be found in [4]. A simplex is a geometrical object in n dimensions that consists of n + 1 points, representing a set of parameter value combinations. In our case of defocus and astigmatism correction n = 3 (astigmatism is adjusted by two controls), and the simplex is a tetrahedron. During every iteration, the Nelder-Mead simplex method first evaluates the sharpness function for a finite number of points (between 1 and n + 2=5). One function evaluation corresponds to one image recording and the sharpness value computation. Based on the function evaluation, the simplex is adapted towards the optimum and will finally encompass the optimum. The last phase will be the simplex closing in on the optimum, until a stopping criterion is reached.

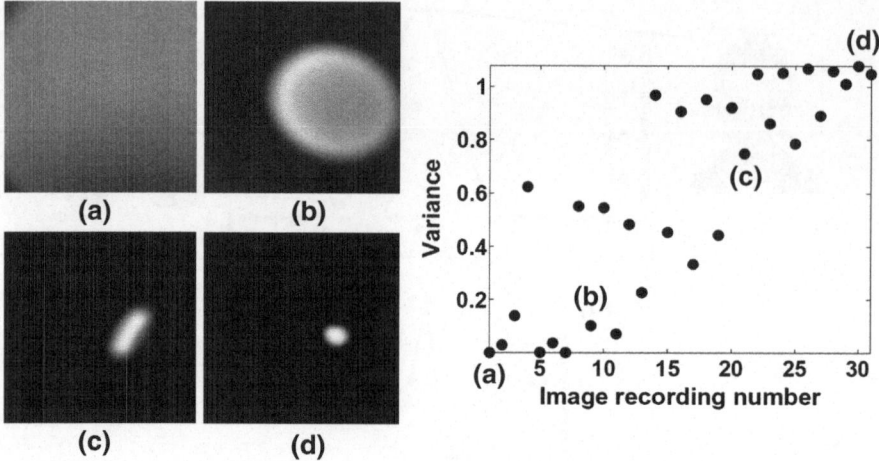

Fig. 6.10 Simultaneous defocus and astigmatism correction for a gold particle with a radius of about 20 nm

Figure 6.10 shows an example of the algorithm's run performed for the out-of-focus stigmatic image of a gold particle with a radius of 20 nm. The figure shows the images of four recordings during optimisation. As mentioned in the beginning of this section, the *maximum* of the variance-based sharpness function corresponds to the optimal image quality. The image quality at the last function evaluation is approximately equivalent to the one obtained by a human operator.

In Fig. 6.11 a typical example is presented that clearly shows the closing in of the three parameters towards a final value during optimisation.

The Fast Focus Correction Method

The fast autofocus algorithm implements the derivative-based approach described in [5, 6]. It has been shown that, after some basic image processing, the derivative-based sharpness function can be accurately approximated with a quadratic polynomial. This implies that after recording at least three images one can find the approximate optimum focus position. This provides the speed improvement in comparison with the method described above. However, to date the fast correction method can be applied only to the one-parameter (defocus only) problem.[1] Note that the method described in the previous subsection was applied to the

---

[1] Note that the astigmatism effect on the image when changing the focal plane in the Z-direction is much less noticeable than the defocusing effect (blurring). This is the reason why only one parameter can be optimised when the system is already close to focus.

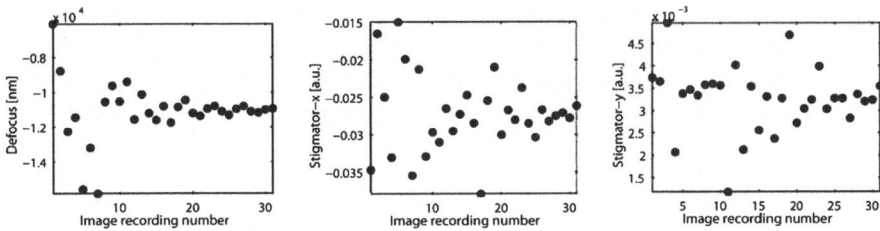

Fig. 6.11 Parameter values of defocus, stigmator-x, and -y versus the recording number, showing the convergence towards the optimum value

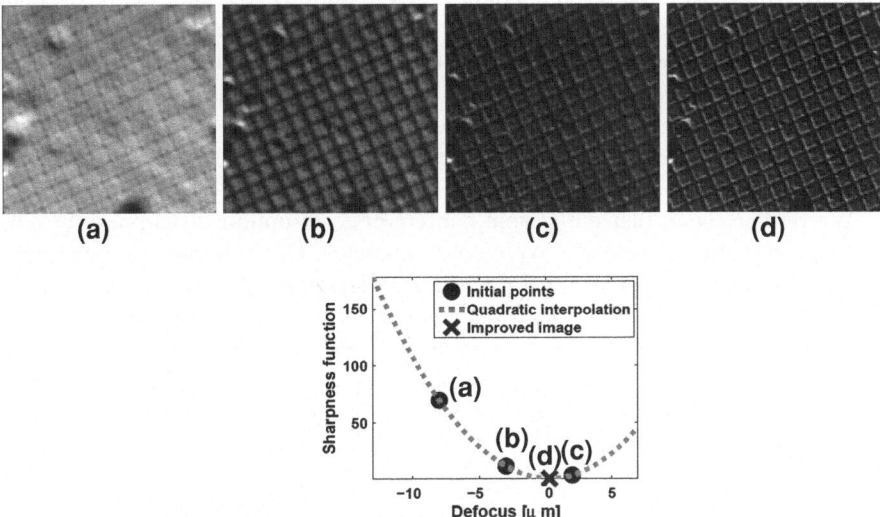

Fig. 6.12 Fast autofocus algorithm example

three-parameter problem (defocus and two stigmators) and can be easily extended to include more parameters.

A fast autofocus algorithm example run is shown in Fig. 6.12 for a carbon cross grating sample. The optimum defocus value of the in-focus image (d) is calculated from the given three images (a, b, c) via the derivative-based sharpness function, interpolated with the quadratic polynomial. This derivative-based sharpness function reaches its *minimum* for the image with optimal sharpness.

Strengths, Weaknesses and Opportunities

The new methods of automated focusing and astigmatism correction described above do not require any special hardware. They are based on sufficiently general assumptions. So they could be considered for different applications, different imaging modes, and for other types of electron microscopes such as SEM, and possibly even for different optical devices.

Though the methods are automated, they do require input parameters, such as the size of initial defocus step and the stopping criterion. We expect that these parameters can be fixed for entire families of sample geometries and for similar microscope settings. This is the case for the particle analysis application used as our guiding use case. Within this application the input parameter values have been chosen experimentally and proved reasonably robust [7].

The simultaneous defocus and astigmatism correction method in general requires about 30–50 image recordings. It takes time comparable to a human operator. Note that the time an operator needs to optimise the microscope depends strongly on experience. An experienced operator stops when knowing very little further improvement can be gained. The method's throughput time can be improved in several ways. They include decreasing the image size, sub-sampling (binning) the image, decreasing the acquisition time, or choosing a faster derivative-free optimisation alternative to the Nelder-Mead simplex method (e.g. the Powell method [8]).

We further found that the system can continue to optimise sharpness beyond what is visible by a reasonably experienced operator. This is because the sharpness function is apparently more sensitive to image sharpness than the human eye.

We successfully applied the methods online at low and medium magnification (see Sect. 3.1). We observed that applying the method in high-resolution mode is not always easy. This is due to disturbances in the environment or microscope instabilities (for instance, image drift and contamination). To apply the optimisation methods routinely in a realistic setting, further research on the robustness of these algorithms is required.

Acknowledgments The author likes to acknowledge Rob van Vucht and Seyno Sluyterman of FEI for their support.

References

1. P.J. van Bree, Control of dynamics and hysteresis in electromagnetic Lenses, Dissertation, Eindhoven University of Technology, 2011
2. M.E. Rudnaya, R.M.M. Mattheij, J.M.L. Maubach, Evaluating sharpness functions for automated scanning electron microscopy. J. Microsc. **240**, 38–49 (2010)
3. M.E. Rudnaya, W. Van Den Broek, R.M.P. Doornbos, R.M.M. Mattheij, J.M.L. Maubach, Autofocus and twofold astigmatism correction in HAADF-STEM. Ultramicroscopy **111**, 1043–1054 (2011)
4. A.R. Conn, K. Scheinberg, L.N. Vicente, *Introduction to Derivative-Free Optimization, MPS-SIAM Series on Optimization* (SIAM, Philadelphia, 2009)
5. M.E. Rudnaya, H.G. Ter Morsche, J.M.L. Maubach, R.M.M. Mattheij, A derivative-based fast autofocus method in electron microscopy. J. Math. Imaging Vision (2011). doi:10.1007/s10851-011-0309-8

6. M.E. Rudnaya, R.M.M. Mattheij, J.M.L. Maubach, H.G. Ter Morsche, Gradient-based sharpness function, in *International Conference of Applied and Engineering Mathematics*, vol. 1, pp. 301–306 (2011)
7. M.E. Rudnaya, Automated Focusing and Astigmatism Correction in Electron Microscope, Dissertation, Eindhoven University of Technology
8. M.E. Rudnaya, S.C. Kho, R.M.M. Mattheij, J.M.L. Maubach, Derivative-free optimization for autofocus and astigmatism correction in electron microscopy, in *2nd International Conference on Engineering Optimization* (2010)

Chapter 7
Positioning Control

Abstract In an electron microscope positioning a sample with respect to the electron beam is performed by a motorized stage or by electromagnetic beam deflectors. At nano-scale many non-linear motion effects occur in the stage, of which the most obvious and annoying effect for an operator is the stick-slip effect. This chapter describes a novel and generic method that uses an optimized control system to provide good results in compensating for this effect. A second disturbing effect is the position drift, i.e. the slow movement of a sample relative to the electron beam. This chapter shows the results of our on-line control approach to compensate for position drift using the two positioning mechanisms. High frequency magnetic and mechanical disturbances that irrecoverably blur the image have to be dealt with in a different way. This chapter also describes an innovative approach to eliminate the adverse effects on the image by fast scanning combined with advanced image processing techniques.

Keywords Position drift · Electron microscopy · Friction compensation · Stick-slip · Drift correction · Control framework · Model-based predictive control · Hierarchical control · Magnetic disturbance · Mechanical disturbance

7.1 Introduction

Richard Doornbos and Sjir van Loo

Embedded Systems Institute, The Netherlands

This chapter deals with positioning a specimen in the electron microscope to get a good quality image of a 'certain area' of the specimen. This is one of the required basic functions of an automated microscope, as discussed in the systems

R. Doornbos and S. van Loo (eds.), *From scientific instrument to industrial machine*,   81
SpringerBriefs in Electrical and Computer Engineering,
DOI: 10.1007/978-94-007-4147-8_7, © The Author(s) 2012

architechure Chap. 2 (Sect. 2.1). For positioning the specimen, three separate problems can be distinguished. First, at low speeds (order tens of nanometres per second) the specimen movement halts and starts again in an irregular manner. This is annoying for an operator when e.g. manually trying to position a feature exactly in the middle of the viewing screen. Secondly, when the stage motors are stopped the stage itself continues to move. After some time the speed reduces but *never* stops. Also various speeds and unpredictable directions of this drift are observable. Currently operators have to wait e.g. 30 min after a stage movement, until the drift is sufficiently small. The third problem can be attributed to the influence of external phenomena like acoustical vibrations or magnetic field changes. These become visible in the apparent position of the specimen in the field of view.

To establish a good understanding of these problems, we first briefly explain positioning in current electron microscope systems. In a typical electron microscope the specimen is moved by a stage, which has up to 5 degrees of freedom: X, Y, Z, roll (also known as 'alpha tilt'), and pitch (also known as 'beta tilt'). Another positioning option is by displacing the electron beam, where two degrees of freedom are possible (X, Y). In this chapter we focus on X and Y positioning.

To get some understanding of the situation in the electron microscope, two figures depicting the two main operational modes are shown. In transmission mode (TEM), a wide parallel beam interacts with the specimen (see Fig. 7.1), therefore only stage movement is relevant in positioning. In scanning transmission mode (STEM) a narrow beam scans across the specimen (Fig. 7.2), therefore stage movement *and* electron beam movement with respect to each other are relevant in positioning.

Given this setting, it is useful to understand which typical movement use cases exist as part of manual and automated operation. The most basic use case is the point-to-point movement. This needs to be a fast move to a next position, and then quickly halting to a full stop. A second use case is the survey scan movement. This is a smooth movement along one dimension, usually along one actuator axis. With this movement one can make a long scan in one direction, intended for survey images. A third important case is the 'stand still', for which the main requirement is to have no movement for a longer time (which proves to be quite difficult). A last movement case is jogging, which needs a smooth movement along a certain path in multiple (possibly 6) dimensions. This last case is often performed by a human operator observing large areas of the specimen, controlling the motion via the stage joystick.

To put positioning in a different perspective we can also look at the various scales of the positioning functions. Note that each scale has different problems and associated solutions.

Notably, when going from the micro- to nano-movements in Table 7.1, a crucial line is crossed. One can consider this to be the well-known paradigm shift in physics when going from the continuum to the discrete world. At nano-scale, motion *is* different! In this regime the discreteness of matter has a strong influence on observable physical properties, which can be thermal, mechanical (frictional and elastic) and kinetic. At nano-scale, we are out of the realm of well-known

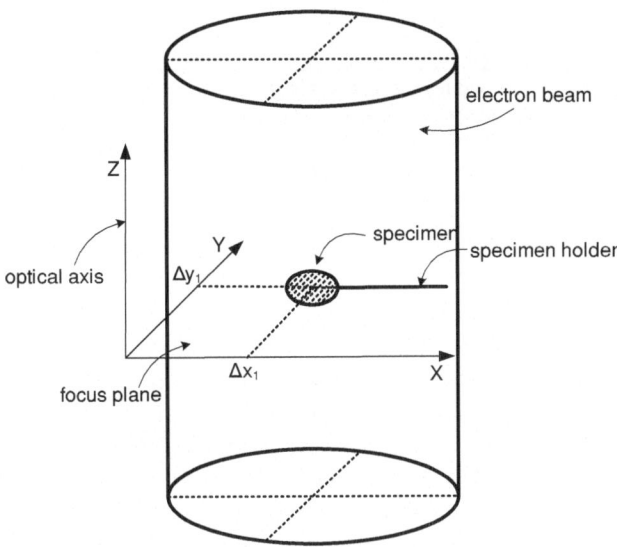

Fig. 7.1 Configuration in TEM mode

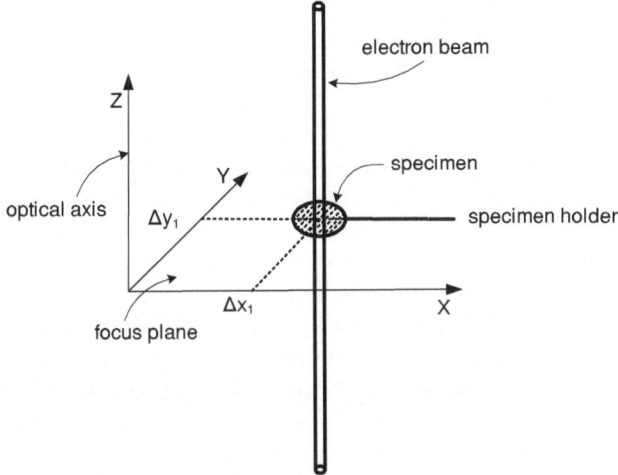

Fig. 7.2 Configuration in STEM mode

'linear' descriptions and have arrived in the complex realm of non-linearities and even quantum effects.

It is clear that this has significant consequences for the positioning functions in electron microscope systems, which encompasses both micro- and nano-scale, and require a smooth transition between both regimes. This impacts both measurability and actuatability.

Table 7.1 Size scales of movement and positioning

Scales of movement	Purposes (examples)	Range
Human scale movements	Loading and unloading the specimen (part of the customer's workflow)	Down to mm
Long range movements	Stage homing, selection of specimen area	Hundreds of μm to mm
Micro-movements	Moving to region of interest	Tens of nm to hundreds of μm
Nano-movements	Drift correction, centering objects	Fractions of nm to tens of nm

Firstly, measurability is different at nano-scale. Measuring properties like position is much more difficult because less matter will generate the signal. The signal is therefore smaller and signal-to-noise ratio closer to the noise floor. In addition, one might say that 'at nano-scale everything moves'.[1] It is clear that in practice it is very hard to define an absolute reference of position at the nano-scale.

Secondly, the actuatability is different at nano-scale. Whereas actuation and movement at the micro-scale (e.g. in the specimen stage) is typically performed using electrical DC or stepper motors, gears, etc., actuation methods at nano-scale are still in development and currently are not yet applied in electron microscopes. The other method of positioning in scanning electron microscopes involves beam deflection; this does not involve displacement of material and as such is of a different nature. Beam deflection is controlled by electromagnetic coils, on which the previous chapter (Chap. 6) elaborates extensively.

Both high quality measurability and actuatability are needed for predictable electron microscopy systems, as the specimen and beam positions are influenced by external and internal disturbances and side-effects. The most prominent external influences are mechanical vibrations (low and high frequency, e.g. from heavy machinery and e.g. from human voices), changes in magnetic fields (e.g. from nearby traffic) and ambient air pressure and temperature changes. These disturbances each have different consequences at different timescales in the system. They have to be sufficiently shielded, as well as actively corrected. Internal disturbances are caused by components in the system itself and are hard to prevent, but are predictable to some extent. We can distinguish two types of effects: those that show in a passive or stationary state, and those that show when the system is active and intended transitions occur. An example of the first type is 'drift', which is visible by image movement over time, although the image is supposed to stand still. Image movement is partly caused by the aforementioned external disturbances, but also by relaxation effects of e.g. the internal temperature distribution (changes cause displacement by thermal expansion), and the mechanical parts of the specimen stage (which behave like discharging loaded springs). Examples of the second type of effects are sideways movements, e.g. during a stage move (sometimes causing the system to go out of optimum

---

[1] At least at temperatures above absolute zero.

focus), and stick–slip (non-smooth) movements. A further complication of movements at nano-scale is backlash, which is caused by clearance between stage gears. It is clear the mechanical system cannot be considered rigid anymore (it is a composition of non-linear springs and dampers). Therefore there will be hidden states that have accumulated energy, which leads to autonomous behaviour (i.e. movements at unpredictable moments).

Clearly, simple actuation to change settings or properties is insufficient when working at nano-scale. A general approach is to find solutions using control theory methods: the measured information about a certain property or state is used to steer the actuators in the desired direction. Sensing the specimen position in the electron microscope is problematic at nano-scale. And it is exacerbated by practical factors such as being unable to locate the sensor at the optimum position (this could block the system's operation), and the very small space available near the optimum position.

Few technology options exist for nanometre-scale sensing of position and movement (e.g. optical interferometry, capacitive sensing). Unfortunately these have an insufficient combination of accuracy and range, or are too costly for the given measurement task. The remaining option to sense displacements is *virtual sensing* using multiple electron microscopic images, i.e. use the system's primary output for sensing purposes. The main disadvantages of this method are (1) reduction of the system throughput, (2) dependency on the microscopy application (and specimen type), and (3) dependency on the image content (which may not have any contrast at all, e.g. when there is a hole in the specimen).

In the following sections, three investigations are discussed to control the positioning in face of the movement issues. Firstly, a solution to reduce the stick–slip effect during the jogging movement is discussed (Sect. 7.2). Secondly, we elaborate on the complexities in controlling a combination of stage and beam movements e.g. for drift compensation (Sect. 7.3). And lastly, we discuss how to compensate for high-frequency disturbances in scanning systems (Sect. 7.4).

7.2 Compensation of Friction in Motion Systems

David Rijlaarsdam and Alina Tarău

Mechanical Engineering, Eindhoven University of Technology, The Netherlands

An operator typically aims to visualise a particular part of a sample by moving it through the electron beam. The motion stage that moves the sample has to move accurately and smoothly to allow visualisation of the sample during motion. The mechanics involved in the stage suffer from friction, which leads to significant performance degradation in the desired smooth movement of the stage (Fig. 7.3). However, despite this undesired effect, that same friction is necessary to keep the stage still.

Fig. 7.3 How hard to push to get it moving?

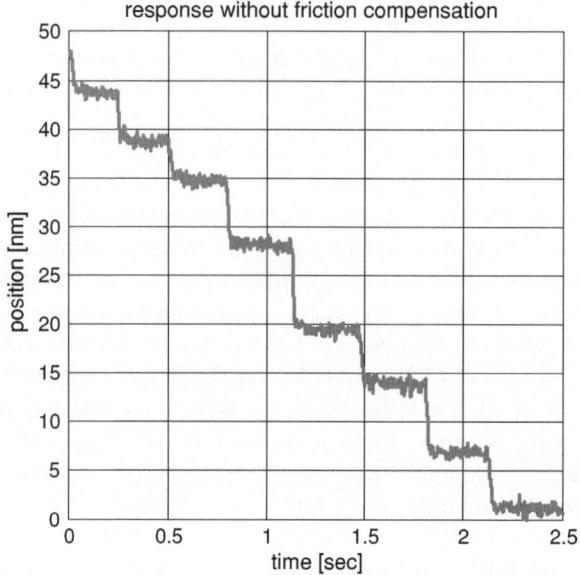

Fig. 7.4 Motion stage in a transmission electron microscope moving at 20 nm/s. The stick–slip induced stepping is unacceptable

Figure 7.4 shows an example of the motion stage position as a function of time. The reference motion in this case is a very slow ramp moving the system at 20 nm/s, a speed typically required when scanning the sample for interesting features. The stick–slip induced stepping shown in Fig. 7.4 is not smooth and therefore unacceptable.

Overcoming this problem requires optimal friction compensation to allow improved microscope operation. In general, non-linear phenomena like friction degrade microscope performance in various use cases. The method introduced in this section allows improved control of such nonlinear effects by applying a relatively simple, but new, engineering approach. Using this method yields a significant performance improvement without requiring advanced nonlinear control algorithms or identification methods, unlike most methods currently used to cope with performance degrading nonlinear effects.

Introduction to the Method

Every real life system is nonlinear to some extent. Although a linear approximation can suffice to model, analyse and control such systems, nonlinear effects pop up when performance requirements increase. Or nonlinear effects may be inherently present in the design. Friction is one example of performance degrading phenomena, but they can also originate in, e.g. nonlinear amplifiers and springs.

Figure 7.4 depicts a typical response when a (proportional) controller aims to position a mass and follow a reference position in the presence of friction. This example shows that the system alternatively sticks and slips, which significantly reduces the system's tracking performance. Hence, the controller is unable to counteract the performance degrading nonlinear effects and alternative methods are required to optimise the system's performance.

The friction in this case can be modelled as Coulomb friction. Although highly simplified, the Coulomb model is often used for friction modelling and control. In Coulomb friction, the nonlinear friction force keeps the system at rest (stick) until the force exceeds the static friction force and the system starts to move (slip).

New: A Frequency Domain Approach

When nonlinear effects are present in a system, the input and output are nonlinearly related. Although this nonlinearity can generally be detected [1, 2] and sometimes modelled [3], this requires considerable effort by the designer. Instead, the method introduced here proposes using a specific excitation signal for detecting nonlinear effects and quantifying/visualising the performance improvement by applying a nonlinear compensator [4, 5].

If a nonlinear system is excited by a sinusoidal (one-tone) input, the output is no longer a sinusoid as it has been deformed nonlinearly. An example of such nonlinear deformation is flattened tops of the sinusoidal signal in Fig. 7.5, due to friction induced stick–slip. Apart from the excitation frequency, the output signal will generally contain other harmonic components (i.e. components having frequencies that are a multiple of the excitation frequency) as well [6]. Although nonlinear phenomena may be visible in the time domain, it remains hard to quantify their effect on the system's performance.

An example output spectrum of a nonlinear system excited with a sinusoidal input is provided in Fig. 7.5. The graph shown in Fig. 7.5 typically results from the nonlinear effect of Coulomb friction. Coulomb friction is an 'odd' nonlinear phenomenon as only odd harmonics show up in the frequency domain. However, the method is not limited to phenomena showing only odd harmonics.

Feed forward controllers compensating for Coulomb friction are well known. Coulomb friction compensation is present in many commercial, off the shelf controller devices. Their general system layout is depicted in the centre of Fig. 7.5, where the goal is to tune the controller parameter K to minimise the performance degrading nonlinear effect, i.e. friction.

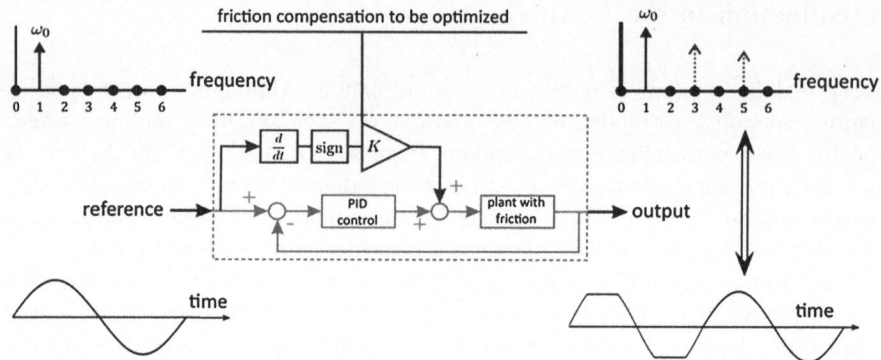

Fig. 7.5 A frequency domain based approach to optimal compensation of performance degrading nonlinear effects

However, tuning the feed forward controller, i.e. the Coulomb friction feed forward gain, for optimal control is very cumbersome. Optimal control implies that the system is optimally linearised. The presence of harmonic components gives a clear and highly sensitive measure for nonlinear effects: *the harmonic components tend to zero as nonlinear effects vanish.* This forms the basis for this new method, as the feed forward controller must be optimised to minimise the performance degrading effects of nonlinearities.

The new method described here introduces a performance cost function which quantifies the combined effect of harmonics induced by nonlinearities (Fig. 7.6), i.e.

$$cost\,function = \sum \frac{amplitude\ at\ relevant\ harmonics}{amplitude\ at\ excitation\ frequency}$$

This cost function is measured as a function of the feed forward controller parameters, in our case the Coulomb friction feed forward gain. Minimising this cost function allows optimal tuning of the feed forward controller parameters to compensate for nonlinear effects, as indicated in Fig. 7.6.

This concept is presented in [7]. Although a number of interesting questions remain, e.g. optimisation of multiple parameter controllers and the solution's generally local nature, our case study proved this method to be very effective. In this case study the method was applied to optimal control of the motion stage stick–slip problem introduced above. The results are discussed in the next subsection.

Application: Optimal Feed Forward Friction Compensation

Friction is often compensated using a simple Coulomb friction model in a feed forward structure. The main issue is accurately tuning the model's friction parameter, i.e. finding the controller parameter in Fig. 7.5. In some cases, careful

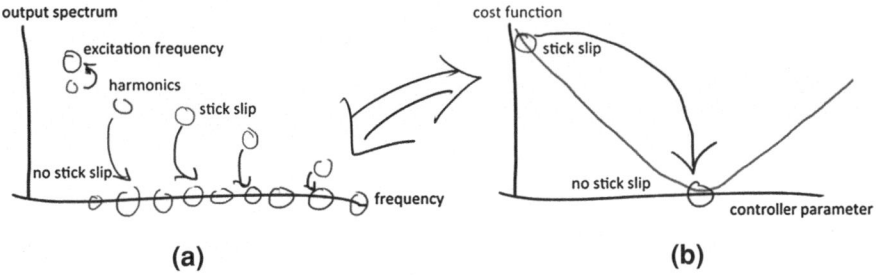

Fig. 7.6 Frequency domain based optimal control of nonlinearities used to minimise friction induced stick–slip

examination of the time domain response yields information to tune such parameters. However, the compensation accuracy required in the electron microscope cannot be obtained this way.

The frequency domain based method is highly sensitive to the presence of nonlinear effects. It allows clear detection of optimal tuning of the feed forward parameters. The result is depicted in Fig. 7.7, where the required speed is again 20 nm/s. Clearly, the friction induced stick–slip effects have been significantly reduced.

Summary and Extensions

In this section, a novel method to analyse, visualise and compensate nonlinear effects in dynamical systems has been introduced. This method allows optimal tuning of controllers for nonlinear systems such that the performance degrading effects of nonlinearities are optimally compensated. It can be easily applied in industry because of the relative simple nature of the excitation signal and a minimum requirement on data analysis and software tools. Highly sensitive, it allows separate analysis of linear and nonlinear effects while providing a clear indication of the best performance that can be obtained using a given control structure. At present the method allows 'one time' tuning only, and time varying effects such as wear and temperature dependent effects cannot be captured. However, if the concept can be extended to adaptive control structures it may be possible to accurately compensate time varying effects as well.

The method is demonstrated in an industrial setting by optimising the performance of an electron microscope's motion stage. Using a simple and widely available control structure presents a significant performance improvement, obtained by optimising the controller parameters. Although the method is demonstrated for optimal friction compensation in motion systems, it should be applicable to a variety of nonlinear control problems. Moreover, automation of the optimisation procedure and extensions towards multi parameter models is

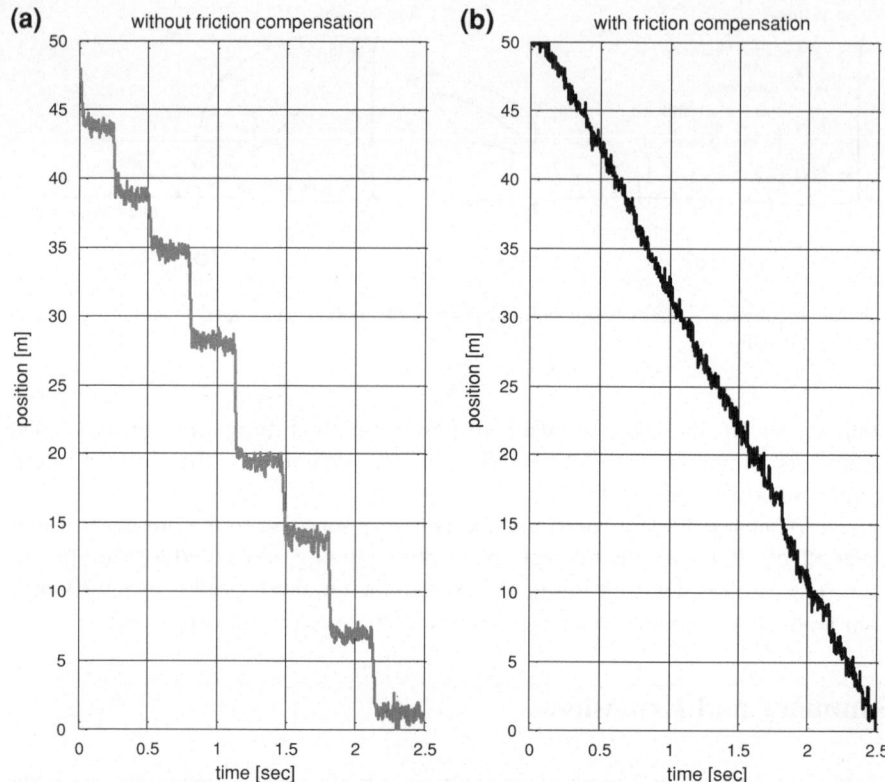

Fig. 7.7 Motion stage in a transmission electron microscope moving at 20 nm/s, without (*left*) and with (*right*) optimal feed forward friction compensation. The optimal parameter setting is obtained using the frequency domain method discussed in this section. It yields a significant improvement in performance as friction induced stick slip is heavily reduced by optimising the control action

expected to yield promising results. This would greatly reduce the efforts required by the user and, e.g. allow easy calibration of industrial systems after maintenance.

Acknowledgments The authors like to acknowledge Twan van den Oetelaar of FEI for his support.

7.3 Drift Correction in Electron Microscopes

Alina Tarău, Pieter Nuij and Maarten Steinbuch

Mechanical Engineering, Eindhoven University of Technology, The Netherlands

Introduction

State-of-the-art electron microscopes perform qualitative evaluation of the internal structure, composition and properties of a wide range of materials as they can show details of samples down to atomic level. At high magnification, when a specimen holder at room temperature is inserted into a microscope, it can take several minutes until a thermal equilibrium is established. During that time the operator can only wait and check periodically if the drift is down to an acceptable level. The situation is worse when thermal processes are induced in a specimen. The specimen can be e.g. cooled to the temperature of liquid nitrogen when looking at biological or soft samples, as this reduces damage caused by the electron beam. Some scientists use heating holders where the specimen can be heated up to 1,000°C while being observed in the electron microscope. These thermal processes involve contraction or expansion of the specimen holder. When using high magnification the image that one looks at moves at high speed. In some cases it may take up to 30 min until the image stabilises. The situation becomes even more severe when gasses are injected in the specimen holding chamber to monitor their reaction on the specimen. Once the cooling/heating is started or the gas injected, the reaction proceeds at its own pace. Therefore, the observations in the microscope have to be performed fast. Thus drift correction is needed.

Approaches for Drift Correction

There are various methods to compensate for image drift, namely image post-processing or on-line compensation schemes. The latter involves active movements of the specimen holder and the electron beam.

Many methods for image post-processing have been proposed in the literature, see e.g. [8–10]. However, for relatively high drift speed with respect to the considered field of view, measurements are blurred and cannot be sharpened using image post-processing only. This makes image post-processing a viable tool for drift correction only when drift speeds are relatively low with respect to the considered field of view.

Several on-line compensation methods have been proposed in the literature for controlling the thermal drift, see e.g. [11]. In these the drift (assumed to be constant over the short term) is compensated for with a series of occasional stage adjustments

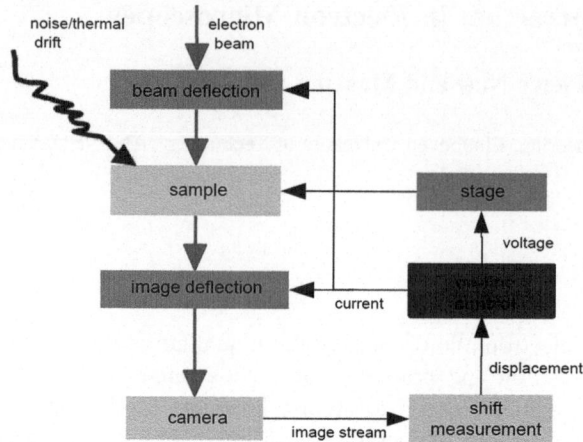

Fig. 7.8 Overall set-up

only. The disadvantage of this method is that for low speed movements, the stage needs to handle the stick–slip friction problems (see previous section).

In this section we describe newly developed and implemented on-line com-' pensation schemes for TEMs only. These schemes involve active movements of the specimen holder, driven by DC motors, and active shifts of the electromagnetic beam deflectors responsible for image deflection while streams of images are acquired from the CCD camera (Fig. 7.8). The goal is to keep the point of interest as close as possible to the point of view at all times. We look at drift with respect to both x and y coordinates. Hence, we deal with a multiple input, multiple output system. Note that the other degrees of freedom of the stage (the z coordinate, α and β) are not addressed in this work.

Next we briefly present the on-line compensation schemes that we developed and implemented. These schemes have different degrees of complexity according to the framework in which they are implemented: centralized or hierarchical, for more details see [12, 13].

Centralised Framework

To obtain the optimal drift correction, from a theoretical perspective, we first proposed a centralised approach. In this a centralised controller computes the voltages used by the DC motors for driving the stage and the currents for the beam deflectors, see Fig. 7.9.

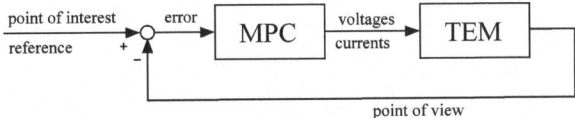

Fig. 7.9 Centralised drift control with respect to x and y coordinates

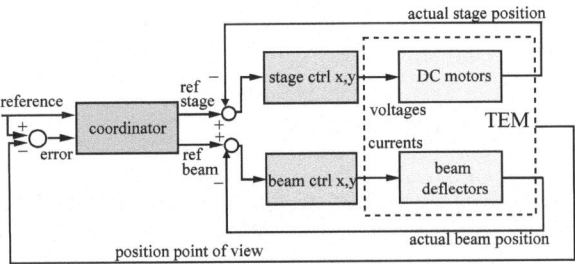

Fig. 7.10 Hierarchical drift control with respect to x and y coordinate

Hierarchical Framework

A less expensive computational approach for drift correction is to develop a drift controller in a hierarchical framework. This framework consists of a multi-level control structure, see Fig. 7.10. It has local controllers at the lowest level and a supervisory controller that coordinates the independent controllers to achieve maximum performance.

The framework layers can be characterised as follows. Given the reference position to be tracked and the actual position (with respect to the x and y coordinate) of the point of view on the image formed on the CCD camera, the coordinating controller decides how much of the reference will be followed by each of the stage and beam controllers. These optimal reference points are then communicated to the local controllers and the microscope adjusts the stage and deflects the beam accordingly.

Furthermore, note that the local independent controllers, that adjust the stage and shift the beam with respect to the x and y coordinate, are developed by using known loop-shaping techniques (see [14] for more details). In this way, we design local PID (Proportional Integral Derivative) controllers that ensure reference point tracking. We consider PID controllers since they can be easily implemented. Also, off-the-shelf components of commercially available motion control systems are available which can be quickly attached to electron microscopes. To make sure that stick–slip problems are reduced, further fine tuning was performed for the stage's x and y controllers. We also determined the optimal feed-forward parameter for Coulomb friction compensation, to reduce stick–slip effects, for more details see [15].

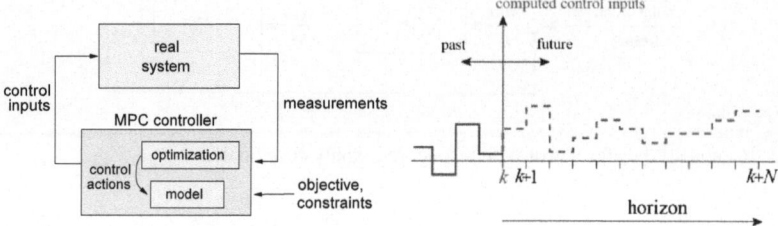

Fig. 7.11 Basic model predictive control

Model-Based Predictive Control

Note that the centralised controller of the first approach and the coordinating controller of the second approach can be efficiently designed as model predictive controllers. The on-line model-based predictive control design method for discrete-time models, see e.g. [16], uses the receding horizon principle.

Basic model predictive control works as follows, see also Fig. 7.11. Given the current state of the system (using measurements), a discrete-time optimisation problem is solved for a system model, over a given prediction period, to achieve a given objective subject to safety and operational constraints. Hence, given a prediction horizon of N steps, at step k (with k the time step counter defined as a positive integer), the future control sequence is computed by solving an optimisation problem. After computing the optimal control sequence, only the first control sample is implemented on the system, and subsequently the horizon is shifted. Next, the system's new state is measured or estimated, and a new optimisation problem at step k + 1 is solved using this new information. In this way, a feedback mechanism is introduced.

Note that due to the feedback mechanism of model predictive control, even if the low-level stage and beam controllers are not designed very efficiently (i.e. they do not track a given reference very accurately), this is automatically corrected by the coordinating controller.

Operational Constraints

The relevant operational constraints derived from the mechanical and design limitations of an electron microscope are the following:

- the voltages applied to the stage DC motors are bounded to avoid damaging the motors,
- the currents applied to the deflection coils are bounded to avoid deflections that would cause the microscope to go out of focus,
- the stage has a limited stroke and the specimen has dimensions that are limited to that stroke.

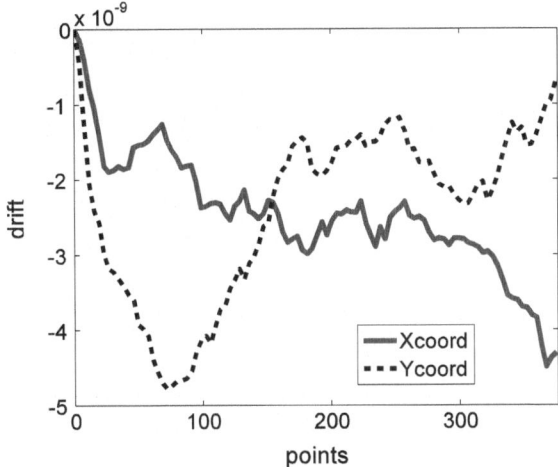

Fig. 7.12 Drift after stage movement

Optimisations

The centralised controller of the first approach and the coordinating controller of
the second approach have to solve nonlinear optimisation problems, [17], for stage
motion models derived in [18] in the model predictive control setting. Typically,
this problem is difficult to solve (NP hard). Therefore, we further considered a
simple and practical approach. Instead of solving a computationally expensive
optimisation problem, we designed efficient heuristic rules for the coordinating
predictive controller [19]. For slow drift these rules involve using beam deflections
only, as long as we do not go out of the deflection bounds. For fast drift we use
stage movements only, while realigning the beam with the optical axis. Finally, we
use a combination of beam deflections and stage movements when we approach
the limits of our constraints.

Results

Here we focus on the hierarchical drift control where at the higher level we
implemented the heuristic coordinator. Simulation results for the centralised and
hierarchical control schemes where we solve nonlinear optimisation problems have
already been published, see [12, 13]. Although the drift compensation schemes
presented above can be used for any type of drift, in this subsection we present the
results obtained when dealing with drift after stage movement only, see Fig. 7.12.
    To predict the drift, we have assumed it is linear on small time intervals (the
sampling frequency should be high enough so that the linearity assumption holds).
We need only two previously detected points to estimate a new one. This method

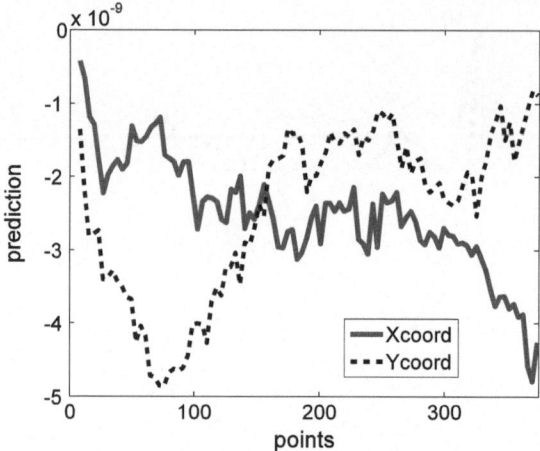

Fig. 7.13 Prediction for drift after stage movement illustrated in Fig. 7.12

has several advantages: it deals with non constant time intervals (often the case when using image sensors); it can efficiently estimate the drift even if we miss recording some points, and the prediction efficiency increases with the sampling frequency. This method performs well even for a relatively low sampling frequency. The results indicate that the error between estimated and realised drift is smaller than 0.5 nm for a sampling frequency of 0.2 Hz. Hence, our assumption of drift linearity is reasonable. Figure 7.13 illustrates the drift prediction plotted in Fig. 7.12 while using the assumption above.

Since the drift after stage movement is relatively slow, we have used only beam deflections to compensate for drift. Hence, as soon as the drift prediction is available, we can smoothly deflect the beam with the estimated displacement. When new images are available, we correct for the remaining errors by deflecting the beam again, according to the new information.

The results for the drift compensation detailed above are illustrated in Fig. 7.14. The remaining errors after drift compensation appear due to the accuracy of drift prediction, accuracy of beam movements, additional noise and hysteresis. However, the errors are typically in the Ångström range and not visible in the image.

Conclusions and Future Work

In this section we proposed an on-line control approach for drift correction in state-of-the-art TEMs. The concept compensates for image drift by actively moving the specimen holder and by shifting the beam deflectors present in these microscopes. Several control approaches and control frameworks are available: centralised/hierarchical control frameworks where the control decisions are determined using

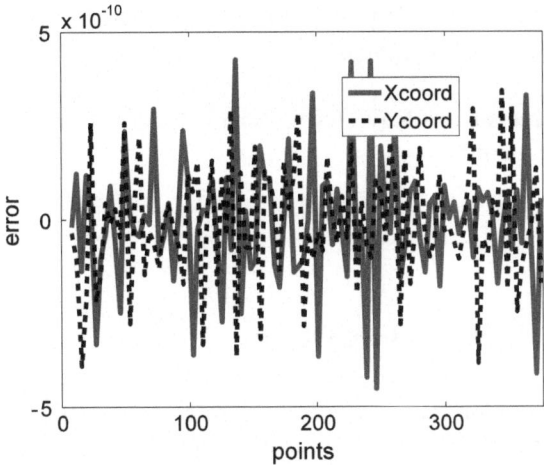

Fig. 7.14 Remaining errors after drift compensation

a predictive control approach. Then optimisation problems have to be solved with different levels of complexity.

Preliminary results show that a hierarchical control framework which fits very well the setting of current electron microscopes can be efficiently used for drift correction: the remaining errors are in the Ångström range.

Finally, the control approaches and frameworks proposed to correct for drift that appears in TEMs can be easily extended to work for other types of electron microscopes, with small changes in the control method due to the different properties of the electron microscopes. Moreover, these control approaches can be successfully used in controlling and optimising other control applications such as transportation systems, e.g. routing problems in traffic control, automated guided vehicles, power distribution and water management, and logistic applications in general.

Acknowledgments   The authors like to acknowledge Edwin Verschueren and Seyno Sluyterman of FEI for their support.

7.4 Compensation of High-Frequency Disturbances in Scanning Microscopes

Pieter Kruit, Vincent van Ravesteijn, Bernd Rieger, Frans Berwald
and Han van der Linden

Applied Sciences, Delft University of Technology, The Netherlands

Introduction: Industrial Problems

Although the 'resolution race in electron microscopy is over', this is only true in
ideal, non-industrial conditions. In practice, many microscopes only reach their
specified resolution after:

- The lab space has been adapted to fulfil the 'pre-installation requirements',
  which sometimes leads to the construction of costly dedicated electron
  microscopy buildings.
- Searching for and eliminating magnetic and mechanical disturbances from other
  machines, elevators, railways and 50/60 Hz power lines.
- Adapting the working hours, i.e. at night or in the weekends, when the environment is quieter.

All these measures characterise scientific instruments, not routine machines.
Our aim is to investigate ways to correct for disturbances, to enable electron
microscopes to be used as routine machines in a realistic industrial environment.

Research Questions

Automation of getting a 'sharp image' is one of the required functions for using
electron microscopes as routine inspection machines. For this, it is necessary to be
able to align, focus and stigmate the beam without operator input. A standard
'auto-focus' method is insufficient to get a sharp image if the beam and sample
move with respect to each other at high frequency. 'Sharpness' not only means
edges go from black to white over a small distance, it also means low image noise
and small distortions. If edges are sharp but ragged from image line to image line,
the user is not happy.

In scanned images, the effects of high-frequency disturbances are noticeable as
displacements of image sections with respect to each other. Most clearly these are
seen on edges in the direction of slow scan (vertical), see Fig. 7.15a and b. Next to
distortion, the second factor influencing image quality is noise. An important
contribution to image noise results from the finite number of electrons per pixel.

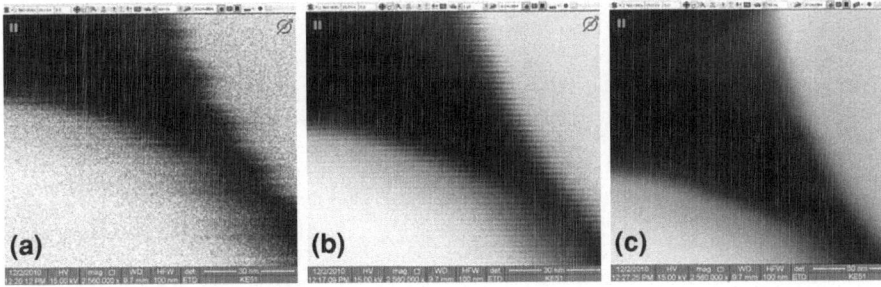

Fig. 7.15 High resolution SEM images (field of view is 120 nm) of small Sn balls. Some drift is visible between the images. **a** Relatively fast scan showing image noise (image graininess) and distortion caused by disturbances (two spatial frequencies are visible, one with a long wavelength and one with a short wavelength). **b** Slower scan showing less image noise and only one spatial frequency distortion (the long wavelength disturbance of Fig. 7.15a has contracted to a short wavelength disturbance here). **c** Averaged image showing less noise but blurred, especially noticeable on the edges

The statistics on this number causes image graininess which makes it harder to recognise the exact location of edges or the shape of low contrast features. Graininess can be reduced by scanning slower, but in a slower scan the ragged edges are more pronounced, more clearly visible when comparing Fig. 7.15b to a. An alternative to slower scanning is to average several images. This results in blurring the image though, as seen in Fig. 15c.

To get a low noise image with sharp edges, we either have to reduce the relative movement of beam and sample, or remove the distortions from the image before the images are averaged and blurred. The first solution has been the approach for the last 50 years, but recently the technology for fast image capture and processing has become available to enable the second approach. The research questions for the second approach are:

- Can we scan fast enough so that disturbances only give distortions, but no loss of information?
- In these fast scans, is there sufficient signal-to-noise to be able to measure the distortions with the required precision?
- Does the correction procedure create image artifacts?
- Can the whole procedure for the correction be performed so fast that the operator only sees the corrected images?

Research Approach

The exact effect of the disturbances on the images was analysed. Most users only see the disturbances from the movement of the image lines in the direction of the fast scan. However, there is also a less easily recognised effect in the direction

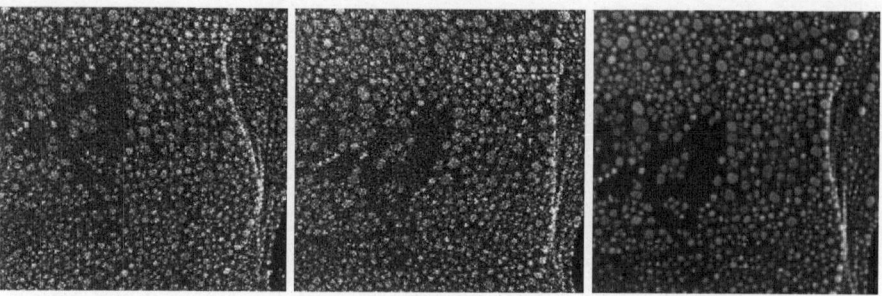

Fig. 7.16 Test of the correction procedure. (**a** and **b**) two out of 32 intentionally distorted SEM images. (**c**) The result of correcting the distortion in the 32 images and averaging

perpendicular to the (horizontal) fast scan direction: the beam moves on and off a specific feature between scan lines. This may result in loss of information and the appearance of ghost features (an example is visible in the left top corner of Fig. 7.15a and b: some white lines, clearly with signal from the Sn ball, are not attached to the rest of the image of the ball).

From this analysis we devised a correction algorithm. The core concept underlying the method is to acquire an image so fast that a significant part of the image (at least 20 image lines) is captured within about 10% of the disturbance period, so there is negligible distortion in this section of the image. In subsequent images of the same object, the same image sections may be shifted with respect to each other. So the algorithm shifts all image sections of the same part of the object back, so they now overlap and can be added without loss of resolution. This means within each corrected image the separate image sections are shifted with respect to each other.

On a side track, we investigated the possibility of using the information on the disturbances to compensate for them in real-time, more specifically, by predicting future disturbances and feeding that information to the deflection coils. This is especially useful for high resolution beam-induced deposition, as the beam is simply positioned more accurately with respect to the sample by this real-time correction method.

Research Results

To test the procedure, a magnetic disturbance was created by running a 50 Hz AC current through coils outside the microscope. Figure 7.16a and b show two of the 32 images that were captured sequentially. Notice both the level of noise in the images and the different distortions, especially visible in the shape of the bright vertical line. Figure 7.16c shows the result of correcting the distortion in the 32 images and averaging the corrected images. Notice that the intrinsic sharpness of the features in Fig. 7.16a and b is not visible because of the image noise. Notice also the increase in apparent contrast in Fig. 7.16c.

Conclusion and Outlook

We have shown that we have found a viable approach to correcting for high-frequency disturbances in scanned images. Although many details have to be investigated further, the results demonstrate that modern image processing (e.g. supported by fast image processing hardware) is a very interesting alternative to traditional correction solutions.

References

1. R. Pintelon, J. Schoukens, *System Identification: A Frequency Domain Approach* (IEEE Press, NY, 2001)
2. J. Schoukens, J. Lataire, R. Pintelon, G. Vandersteen, T. Dobrowiecki, Robustness issues of the best linear approximation of a nonlinear system. IEEE Trans. Instrum. Meas. **58**, 1737–1745 (2009). doi:10.1109/TIM.2009.2012948
3. D.J. Rijlaarsdam, P.W.J.M. Nuij, J. Schoukens, M. Steinbuch, Spectral analysis of block structured nonlinear systems and higher order sinusoidal input describing functions. Automatica **47**(12), 2684–2688 (2011). doi:10.1016/j.automatica.2011.08.049
4. D.J. Rijlaarsdam, P.W.J.M. Nuij, J. Schoukens, M. Steinbuch, Frequency domain based friction compensation, industrial application to transmission electron microscopes, in *Proceedings of American Control Conference*, pp. 4093–4098 (2011)
5. D.J. Rijlaarsdam, P.W.J.M. Nuij, J. Schoukens, M. Steinbuch, Frequency domain based nonlinear feed forward control design for friction compensation. Mech. Syst. Signal Process. **27**, 551–562 (2012). doi:10.1016/j.ymssp.2011.08.008
6. P.W.J.M. Nuij, O.H. Bosgra, M. Steinbuch, Higher-order sinusoidal input describing functions for the analysis of non-linear systems with harmonic responses. Mech. Syst. Signal Process. **20**, 1883–1904 (2006). doi:10.1016/j.ymssp.2005.04.006
7. D.J. Rijlaarsdam, V. Van Geffen, P.W.J.M. Nuij, J. Schoukens, M. Steinbuch, Frequency domain based feed forward tuning for friction compensation, in *ASPE Spring Topical Meeting*, pp. 129–134 (2010)
8. R.C. Gonzalez, R.E. Woods, *Digital Image Processing* (Addison-Wesley, Boston, 1992)
9. B.S. Salmons, D.R. Katz, M.L. Trawick, Correction of distortion due to thermal drift in scanning probe microscopy. Ultramicroscopy **1**(3), 339–349 (2010)
10. M.A. Sutton, J.J. Orteu, H. Schreier, *Image Correlation for Shape, Motion and Deformation Measurements: Basic Concepts, Theory and Applications* (Springer-Verlag, NY, 2009)
11. M.A. O'Keefe, B. Parvin, D. Owen, J. Taylor, K.H. Westmacott, W. Johnston, U. Dahmen, Automation for on-line remote-control in situ electron microscopy. Scanning Microsc. **11**, 229–239 (1997)
12. A.N. Tarău, P.W.J.M. Nuij, J. Schoukens, M. Steinbuch, Hierarchical control for drift correction in transmission electron microscopes, in *Proceedings of the 19th IEEE International Conference on Control Applications*, (2011) pp. 351–356. doi: 10.1109/CCA.2011.6044492

13. A.N. Tarău, P.W.J.M. Nuij, M. Steinbuch, Model-based drift control for electron microscopes, in *Proceedings of the 18th IFAC World Congress*, (2011), pp. 8583–8588. doi: 10.3182/20110828-6-IT-1002.02190

14. S. Skogestad, I. Postlethwaite, *Multivariable Feedback Control: Analysis And Design* (Wiley, Chichester, 2005)

15. D.J. Rijlaarsdam, P.W.J.M. Nuij, J. Schoukens, M. Steinbuch, Frequency domain based nonlinear feed forward control design for friction compensation. Mech. Syst. Signal Process. **27**, 551–562 (2012). doi:10.1016/j.ymssp.2011.08.008

16. J.B. Rawlings, D.Q. Mayne, *Model Predictive Control: Theory and Design* (Nob Hill Publishing, Madison, 2009)

17. P.M. Pardalos, M.G.C. Resende, *Handbook of Applied Optimization* (Oxford University Press, Oxford, 2002)

18. D. Rijlaarsdam, B. Van Loon, P. Nuij, M. Steinbuch, Nonlinearities in industrial motion stages–detection and classification, in *Proceedings of the American Control Conference*, pp. 6644–6649

19. A.N. Tarău, P.W.J.M. Nuij, M. Steinbuch, Practical approach towards drift correction in electron microscopes, in *Technical Report, Control Systems Technology* (Eindhoven University of Technology, 2011)

Part IV
Conclusion

Chapter 8
Final Words

Architectural Stress

This book describes our efforts related to the difficulty in adapting a well-established system, like the electron microscope, to new uses and functions as demanded by new markets, or by changes in the current markets. We call this inherent and designed-in inability to respond to changes 'architectural stress'.

Stress in the architecture appears in general when a system has been so optimised and specialised for a single type of function or a single set of qualities, that it cannot be adapted easily to meet a change of required function or qualities, usually determined by an external force (e.g. a change in market demands).

In our case the stress is caused by our desire to make rapid measurements using a scientific instrument that is not designed to be fast. Rather, it is optimised for ultimate resolution and high image quality, both realised by inherent stability mechanisms which take significant time. Furthermore, due to this strong optimisation, little margin is left for adjustment and adaptation of the system parameters to compensate for adverse system behaviour.

Another source of stress is our desire to automate measurements using a scientific instrument traditionally intended for human operators. Human operators use manual controls and use their eyes (and brains!) to set the system optimally. This explains why FEI microscopes were not sufficiently instrumented for our experimental purposes.

Our desire to obtain accurate and precise measurements using a scientific instrument not designed for absolute accuracy also leads to architectural stress. Accuracy, i.e. yielding correct values with respect to external standards, can be achieved by calibrating the system. Calibrations can be made in principle, but require external reference samples or are based on atom distances, both unfeasible for the new type of microscopy applications. Furthermore, the system should have an accuracy notion, and be able to maintain a certain accuracy degree over its operational states. Precision, the ability to reproduce the same results for the same

R. Doornbos and S. van Loo (eds.), *From scientific instrument to industrial machine*,   105
SpringerBriefs in Electrical and Computer Engineering,
DOI: 10.1007/978-94-007-4147-8_8, © The Author(s) 2012

inputs, has never been the most important design driver for the electron micro-scope's architecture. Therefore improvements here may also be needed.

Coping with Architectural Stress

Essentially we are stuck in a situation where we want to achieve something with the system, but the system does not allow it easily. The main question is: how do we act to relieve the architectural stress to achieve a system that does what we want? What should be changed, what can be changed, and how?

The Project

Because it was a research project we were unable to, not allowed to, or it was infeasible to change major parts of the electron microscope's hardware and mechanical design. We therefore restricted ourselves mainly to the electronic and software subsystems, leading to the questions of what can we do for this issue using steering and control.

One can distinguish two approaches to counter the stress: a fundamental approach via the discipline axis, or a multi-disciplinary systems approach. In our project, both approaches were used to investigate which fundamental changes needed to be addressed in developing a new architecture.

In the first place, at engineering level, the focus was on speeding up physical transitions, e.g. lens setting (Sect. 6.2) and stage moves (Sect. 7.2). Gathering insights into factors related to speed determined by software (Sect. 4.1) also fitted into this category.

Secondly, at system level, the focus was on the overall throughput. Here the investigations were guided by a challenging use case, where several contributions from different disciplines addressed a small part of the puzzle. In our feasibility prototype, the Concept Car (Sect. 3.1), these were combined, and together delivered more than initially expected. We investigated many aspects of the system including speed, automation and optimization within the given constraints, and created various new solutions, and ways of thinking and reasoning (Sects. 3.1–3.3, 5.2, 6.3, 7.3, 7.4). However, getting these new ideas accepted for new products depends on business decisions. With respect to changes in the architecture, this is a long term matter, as the systems architecture is only very rarely redesigned or refactored.

The Organisation

At the organisational level, architectural stress was also visible. The way of working and the organisation's culture is entirely focused on realising the qualities put forward by the traditional key drivers, namely ultimate resolution and image quality. Changing key drivers is a major shift for an organisation. Especially as these devices have been around since the 1950s and have a very long product lifecycle. Also, expert knowledge and considerable design experience is essential. All of this is true for the electron microscopy field. It will require a large investment, true management commitment and a very long time before this change is actually realised.

Project Experiences and Observations

Looking back, one of the main issues that emerged during the project execution was the lack of flexibility in the project's direction. The company's expectations and attention changed due to growing insight, which has lead to shorter term topics. For the project to provide as much value for the company as possible, it should have responded to these changes faster, changing direction and its associated research topics appropriately. We clearly identified the need for a more flexible project set-up, as the 4/5 year period was too long.

Another, more general topic is the well-known tension between engineering and science. We experienced this tension in our project in many ways, and it is clearly explained by the difference in interests and goals (including the pay off). The following exaggerated description of typical scientists and engineers makes things clear. Scientists are mainly concerned with creating publications. They select research topics on the chance of success, the ability to make a mathematical proof, and limit the problem's complexity for dissemination reasons. Successfully applying research results in an industrial environment has a lower priority. On the other hand, engineers are mainly concerned with solving immediate problems. They are usually less interested in the scientific or fundamental background of a solution. The focus is on robustness, quality, speed and cost of applying a certain solution.

Closing this gap is a huge challenge but it is worthwhile, as fresh and independent views on an industrial problem generally generate new insights and unexpected solutions. One possible way to address this issue is to put the scientist and engineer in close contact and somehow create a shared responsibility for each other's results.

Worldwide Trends

Architectural stress is more often present in systems than realised. When we look at other product manufacturers we clearly see signs of architectural stress. Many systems have difficulty in responding to changes in use, key qualities or cost demands.

We observed that these changes are accompanied by worldwide trends. They include embedding the product or system in the 'Internet of Things', embedding the system into the customer's workflow (and possibly being a partner in the customer's business), and increased data fusion and multi-modality, which exploits the computation-powered extraction of information. We see all these trends happening, also from the perspective of FEI, and in many cases the consequences are considered to be very important.

Hopefully the ideas, approaches and solutions described in this book may help engineers and architects in similar situations to relieve the architectural stress in their systems.

Appendix
Condor Project Publications

2008

1. P.J. Van Bree, Dynamics of magnetic electron lenses, in *Proceedings of the 27th Benelux meeting on Systems and Control*, (2008), pp. 32–32

2009

1. N. Muhammad, Y. Vandewoude, Y. Berbers, S. Van Loo, Modeling composite end-to-end flows with AADL, in *Proceedings of STANDRTS Workshop on Euromicro Conference on Real-Time Systems*
2. D. Langsweirdt, Y. Vandewoude, Y. Berbers, Towards intelligent tool-support for AADL based modeling of embedded systems, in *Proceedings of the 2nd International Workshop on Model Based Architecting and Construction of Embedded Systems* , vol. 507, (2009), pp. 81–85
3. W. Paszke, O. Bachelier, New robust stability and stabilization conditions for linear repetitive processes, in *Proceedings of 6th International Workshop on Multidimensional (nD) Systems* (2009). ISBN: 978-1-4244-2798-7
4. W. Paszke, P. Rapisarda, E. Rogers, M. Steinbuch, Dissipative stability theory for linear repetitive processes with application in iterative learning control, in *Proceedings of Symposium on Learning Control at IEEE CDC* (2009)
5. W. Paszke, Iterative learning control by Linear Repetitive Processes Theory, in *Proceedings of the 28th Benelux Meeting on Systems and Control*, (2009), pp. 149–150
6. A. Tejada, W. Van Den Broek, S. Van Der Hoeven, A.J. Den Dekker, Towards STEM control: modeling framework and development of a sensor for defocus control, in *Proceedings of 48th IEEE Conference on Decision and Control*, (2009), pp. 8310–8315
7. M.E. Rudnaya, R.M.M. Mattheij, J.M.L. Maubach, Evaluating sharpness functions for automated scanning electron microscopy. J. Microsc. **240**, 38–49 (2009)

R. Doornbos and S. van Loo (eds.), *From scientific instrument to industrial machine*, 109
SpringerBriefs in Electrical and Computer Engineering,
DOI: 10.1007/978-94-007-4147-8, © The Author(s) 2012

8. M.E. Rudnaya, R.M.M. Mattheij, J.M.L. Maubach, Iterative autofocus algorithms for scanning electron microscopy. Microsc. Microanal. **15**, 1108–1109 (2009)
9. M.E. Rudnaya, J.M.L. Maubach, R.M.M. Mattheij, Scanning Electron Microscopy: Power Spectrum Analysis, in *Proceedings of Microscopy Conference* (2009), ISBN: 978-3-85125-062-6
10. A.J. Tejada Ruiz, S.W. Van Der Hoeven, A.J. Den Dekker, P.M.J. Van den Hof, Towards automatic control of scanning transmission electron microscopes, in *Proceedings IEEE Multi-Conference on Systems and Control* (2009), pp. 788–793
11. A.J. Tejada Ruiz, Towards automatic control of scanning transmission electron microscopes: system identification issues, in *Proceedings 28th Benelux Meeting on Systems and Control*
12. A.J. Tejada Ruiz, A.J. Den Dekker, S.W. Van Der Hoeven, Observer development for automatic STEM closed-control loop. Microsc. Microanal. **15**, 1096–1097 (2009). doi:10.1017/S1431927609095336
13. P.J. Van Bree, C.M.M. Van Lierop, P.P.J. Van Den Bosch, Control-oriented hysteresis models for magnetic electron lenses. IEEE Trans. Magnet. pp. 5235–5238
14. P.J. Van Bree, C.M.M. Van Lierop, P.P.J. Van Den Bosch, Characterization of hysteresis within magnetic electron lenses, in *Proceedings 28th Benelux Meeting on Systems and Control*, pp. 81–82
15. W. Van Den Broek, S. Van Aert, D. Van Dyck, A model based atomic resolution tomographic algorithm. Ultramicroscopy **109**, 1485–1490 (2009)
16. W. Van Den Broek, S. Van Aert, D. Van Dyck, Model based tomography in high resolution HAADF STEM, in *Proceedings of Microscopy Conference Graz*, (2009) pp. 95–96. ISBN: 978-3-85125-062-6
17. S.W. Van Der Hoeven, A.J. Den Dekker, A.J. Tejada Ruiz, Alignment control of STEM: a Ronchigram based approach. Microsc. Microanal. **15**, 118–119 (2009). doi:10.1017/S1431927609094501

2010

1. D. Langsweirdt, N. Boucké, Y. Berbers, Architecture-Driven Development of embedded systems with ACOL, in *13th IEEE International Symposium on Object/Component/Service-Oriented Real-Time Distributed Computing Workshops* (2010). doi: 10.1109/ISORCW.2010.2
2. N. Muhammad, Y. Vandewoude, Y. Berbers, S. Van Loo, *Modeling Embedded Systems with AADL: A Practical Study*. New Advanced Technologies (IN-TECH publishers, 2010)
3. N. Muhammad, N. Boucké, Y. Berbers, Using the parallelism viewpoint to optimize the use of threads in parallelism-intensive software systems, in *IEEE International Conference on Software and Computing Technology* (ICSCT 2010)

4. N. Muhammad, N. Boucké, Y. Berbers, A parallelism viewpoint to analyze performance bottlenecks of parallelism-intensive software systems, in *6th Central and Eastern European Software Engineering Conference* (2010)

5. N. Muhammad, N. Boucké, Y. Berbers, Model-based enhancement of software performance for precision critical systems, in *ECSA 2010 Doctoral Symposium* (Copenhagen, 2010)

6. M.E. Rudnaya, S.C. Kho, R.M.M. Mattheij, J.M.L. Maubach, Derivative-free optimization for autofocus and astigmatism correction in electron microscopy, in *2nd International Conference on Engineering Optimization*

7. M.E. Rudnaya, R.M.M. Mattheij, J.M.L. Maubach, Derivative-based image quality measure. CASA-Report (Eindhoven University of Technology, 2010), pp. 10–42

8. M.E. Rudnaya, R.M.M. Mattheij, J.M.L. Maubach, Evaluating sharpness functions for automated scanning electron microscopy. J. Microsc. **240**, 38–49 (2010)

9. D. Rijlaarsdam, B. Van Loon, P. Nuij, M. Steinbuch, Nonlinearities in industrial motion stages-detection and classification, in *Proceedings of the American Control Conference*, pp. 6644–6649 (2010)

10. D.J. Rijlaarsdam, V. Van Geffen, P.W.J.M. Nuij, J. Schoukens, M. Steinbuch, Frequency domain based feed forward tuning for friction compensation, in *ASPE Spring Topical Meeting*, pp. 129–134 (2010)

11. P.J. Van Bree, C.M.M. Van Lierop, P.P.J. Van den Bosch, On hysteresis in magnetic lenses of electron microscopes, in *IEEE International Symposium on Industrial Electronics* (ISIE-2010)

12. P.J. Van Bree, C.M.M. Van Lierop, P.P.J. Van den Bosch, Electron microscopy experiments concerning hysteresis in the magnetic lens system, in *IEEE Multi-conference on Systems and Control, Yokohama* (2010)

13. P.J. Van Bree, C.M.M. Van Lierop, P.P.J. Van den Bosch, Feed forward initialization of hysteretic systems, in *IEEE Conference on Decision and Control* (2010)

14. W. Van Den Broek, S. Van Aert, D. Van Dyck, A model based reconstruction technique for depth sectioning with scanning transmission electron microscopy. Ultramicroscopy **110**, 548–554. doi: 10.1016/j.ultramic.2009.09.008 (2010)

15. W. Van Den Broek, S. Van Aert, P. Goos, D. Van Dyck, Throughput maximization of particle radius measurements through balancing size versus current of the electron probe. Ultramicroscopy **111**, 940–947 (2010). doi: 10.1016/j.ultramic.2010.11.025

16. W. Van Den Broek, S. Van Aert, D. Van Dyck, Model based atomic resolution tomography, in *Abstracts of the 17th International Microscopy Congress* (2010)

17. N. Van Lierop, Phenomenological modeling pitfalls, in *Embedded Systems Institute symposium* (Eindhoven, 2010)

2011

1. B. Goris, S. Bals, W. Van den Broek, J. Verbeeck, G. Van Tendeloo, Exploring different inelastic projection mechanisms for electron tomography. Ultramicroscopy **111**, 1262–1267 (2011)
2. A. Malfliet, W. Van den Broek, F. Chassagne, J.D.Mithieux, B. Blanpain, P. Wollants, Fe3Nb3N precipitates of the Fe3W3C type in Nb stabilized ferritic stainless steel. J. Alloys Compd. (2011) doi:10.1016/j.jallcom.2011.03.155
3. H.H. Mezerji, W. Van den Broek, S. Bals, A practical method to determine the effective resolution in incoherent experimental electron tomography. Ultramicroscopy **111**, 330–336 (2011)
4. N. Muhammad, Architecture level modelling and analysis support for software performance, Dissertation, Katholieke Universiteit Leuven, 2011
5. M.E. Rudnaya, W. Van den Broek, R.M.P. Doornbos, R.M.M. Mattheij, J.M.L. Maubach, Defocus and two-fold astigmatism correction in HAADF-STEM. CASA-Report 10-09, Eindhoven University of Technology, 2010
6. M. Rudnaya, Automated Focusing and Astigmatism Correction in Electron microscopy, Dissertation, Eindhoven University of Technology, 2011
7. A. Tejada, A.J. Den Dekker, W. Van den Broek, Introducing measure-by-wire, the systematic use of systems and control theory in transmission electron microscopy. Ultramicroscopy **111**,1581–1591 (2011)
8. A. Tejada, J.R. Chavez-Fuentes, P. Vos, Stability and performance analysis of dual-random-rate systems via Markov jump linear system theory, in *50th IEEE Conference on Decision and Control and European Control Conference*, (2011)
9. A. Tejada, A.J. Den Dekker, POEM: defocus polar rose estimation method—a fast defocus estimation method for STEM, in *Proceedings of IEEE International Instrumentation and Measurement Technology Conference*, pp. 1228–1232 (2011)
10. A. Tejada, The role of Poisson's binomial distribution in the analysis of TEM images. Ultramicroscopy **111**, 1553–1556 (2011)
11. A. Tejada, P. Vos, A.J. Den Dekker, Towards an adaptive minimum variance control scheme for specimen drift compensation in transmission electron microscopes, in *Proceedings of 7th International Workshop on Multidimensional (nD) Systems*, pp. 1–6
12. P. Van Bree, Control of Dynamics and Hysteresis in Electromagnetic Lenses, Dissertation, Eindhoven University of Technology, 2010